Fiber Optic Installations

Other McGraw-Hill Communications Books of Interest

Fiber Optic Installations

A Practical Guide

Bob Chomycz, P.Eng.

McGraw-Hill

New York   San Francisco   Washington, D.C.   Auckland   Bogotá
Caracas   Lisbon   London   Madrid   Mexico City   Milan
Montreal   New Delhi   San Juan   Singapore
Sydney   Tokyo   Toronto

Library of Congress Cataloging-in-Publication Data

Chomycz, Bob.
    Fiber optic installations / a practical guide / B. Chomycz.
       p.   cm.
    Includes index.
    ISBN 0-07-011635-0
    1. Optical communications—Equipment and supplies. 2. Fiber
optics—Equipment and supplies. I. Title.
TK5103.59.C52   1996
    621.382'75—dc20   95-48304
                                                     CIP

McGraw-Hill

*A Division of The **McGraw·Hill** Companies*

1 2 3 4 5 6 7 8 9 0 DOC/DOC 9 0 1 0 9 8 7 6

ISBN 0-07-011635-0

The sponsoring editor for this book was Steve Chapman, the editing supervisor was Fred Bernardi, and the production supervisor was Pamela Pelton. It was set in Century Schoolbook by Dina E. John of McGraw-Hill's Professional Book Group composition unit.

Printed and bound by R. R. Donnelley & Sons Company.

This book is printed on recycled, acid-free paper containing a minimum of 50% recycled, de-inked fiber.

McGraw-Hill books are available at special quantity discounts to use as premiums and sales promotions, or for use in corporate training programs. For more information, please write to the Director of Special Sales, McGraw-Hill, 11 West 19th Street, New York, NY 10011. Or contact your local bookstore.

Contents

Preface

The primary objective of this book is to present the principles of fiber optic communication and the techniques used to install and commission systems in various communications environments. The emphasis is on practical installation methods using industry standards. The book is written for individuals with little or no knowledge of the field.

The approach I have chosen for this book is not highly theoretical nor do I attempt to cover the physics of lightwave communication. There are a number of good books available that discuss the theoretical aspects of fiber optics, but very few books deal with the practical aspects of fiber optics in the real world. Because of the unique nature of this medium, conventional electrical wire implementation techniques are not applicable. Employees working with fiber optics need to understand not only the basic theory but the practical methods used to implement this technology. This book attempts to accomplish that task.

Fiber Optic Installations: A Practical Guide will be of primary interest to the following professions: technicians, electricians, cable installers, testers, telecommunications personnel, data communications personnel, engineers, technologists, tradespeople, students, and others with an interest in the field.

The procedures and information presented in this book are intended as a general guide only. The assistance of qualified professionals should be obtained when implementing any fiber optic system.

Bob Chomycz

Fiber Optic Installations

Introduction

1.1 The Fiber Optic Revolution

Over the last twenty years, a quiet revolution has been changing the world of communications. Indirectly, it will affect all our lives and enhance our ability to communicate large amounts of information over vast distances with extreme clarity and reliability. This revolution centers on the replacement of existing copper wire communication cables with thin strands of glass fibers that carry light pulses.

Since the earliest recorded history, light has been used to communicate over distances, although the techniques employed have often been slow and cumbersome. Historically, communication has been limited by atmospheric conditions—usually impossible in fog or heavy rain, for example—and restricted to line-of-site operation. As early as the ancient Greeks and Phoenicians, sunlight was reflected off mirrors for signaling between towers, and this technique continued into the modern era with several variations. Eventually, sunlight was replaced by artificial light, and the on/off signaling became more structured until it resembled a Morse code. The military, for example, still uses a version of this technique for low-speed communication between ships.

In the late nineteenth century, Alexander Graham Bell worked on a design for a "Photophone," for sending voice over a light beam. Sunlight was reflected off a mirror that vibrated to voice sound waves. The receiver was a photocell connected to an electric current that passed to a speaker. The idea was good, but the technology was not yet in place to use it practically.

After the laser was invented in 1958, further studies were performed with light communication in air. Lasers provided a narrow band of light radiation that could be bent with mirrors. Communication by light was not practical, however, because it required a clear line-of-sight; fog or rain would quickly obstruct the link. Experiments continued with light propagation in a glass medium. Glass was preferred over air because of its constant nature and because it was unaffected by environmental variations.

In 1970, the first low-loss optical fiber was developed. The optical fiber, made from silica glass about 250 micrometers (μm) in diameter—about the size of a human hair—was used to propagate light in a lab environment. This was the beginning of fiber optics.

Shortly after this, the process of manufacturing thin glass fiber strands was perfected and, in the mid 1970s, Corning Inc. made fiber optic cable available commercially. This launched the fiber optic revolution. Short-distance systems were subsequently tested by many telephone companies. With the continual refinement of the technology, communication distances increased and more products became available.

In 1980, Bell announced the installation of 611 miles of optical fiber in its northeast U.S. corridor. Likewise, Saskatchewan Telephone announced installation of 3600 km of optical fiber in Canada. At the 1980 Lake Placid Winter Olympics, fiber optics was first used to transmit television signals. In the years that followed, fiber optics gradually gained popularity in the telecommunications world. Today, it is a widely accepted and proven technology. The replacement of old wire communication lines with new fiber optic cable is now the norm for many applications.

The basic principle of the on/off light communication used in the past is similar to the principle used in fiber optics today. The information signal to be transmitted controls a light source by turning it on and off in a particular coded sequence or by varying its intensity. The light is then coupled into an optical fiber that guides it for the distance of the communication. At the receiving end, a detector decodes the light and reproduces the information signal.

Although light travels in a straight line in free space, the glass properties of the optical fiber guide the light around bends and allow fiber optic cable routes to work like standard copper wire cable routes, with some restrictions. The distance of propagation is determined mainly by the loss of the light in the optical fiber and by the rate of the on/off signaling. At the other end of the optical fiber the light is coupled to a light-sensitive photodetector, which converts the pulsing light signal back to a usable electrical signal.

Today's fiber optic technology can support transmission rates of over 2 billion light pulses per second. This translates to over 60,000 simultaneous telephone calls. A standard 200-fiber cable can carry over 6 million telephone conversations compared to the 10,000 conversations a similar-sized copper cable can carry.

Optical fiber doesn't care what type of signal it is propagating. This makes it a versatile medium, available for practically all types of communication, including telephone, video, television, images, computers, local area networks (LANs), wide area networks (WANs), control systems, and so on. One fiber optic cable installation can be used for many applications.

As this lightwave revolution matures, we can expect better and more abundant service for our everyday needs. Already, many of our phone calls use fiber optic facilities. Cable TV companies are adding fiber optics to their networks so we can enjoy a larger selection of channels with better quality. Interactive television, which uses fiber optics, is being tested in many locations. Fiber optics is also used in many industries to carry high-speed computer data throughout a factory or across countries.

1.2 Basic Transmission

Fiber optics involves the transmission of information by light through long transparent fibers made from glass or plastic. A light source *modulates* [a light-emitting diode (LED) or laser turns on or off or varies in intensity] in a manner that represents the electrical information input signal. The modulating light is coupled to an optical fiber that propagates the light. An optical detector at the opposite end of the fiber receives the modulating light and converts it back to an electrical signal identical to the input signal.

Light transmission techniques can be divided into three major categories: digital modulation, analog modulation, and digital modulation with analog-to-digital conversion. Digital modulation involves the conversion of the electrical digital input signal into a similar coded sequence of on or off (digital) light pulses (see Fig. 1.1a). Because all computer communications use electrical digital communications, this type of modulation is well suited for computer data transmission.

Analog communication signals, such as for voice or video transmission, vary in electrical amplitude and period. Analog modulation converts this electrical input signal into an optical signal of similarly varying light intensity (see Fig. 1.1b). This technique can be relatively inexpensive and is often used in fiber optic modem applications.

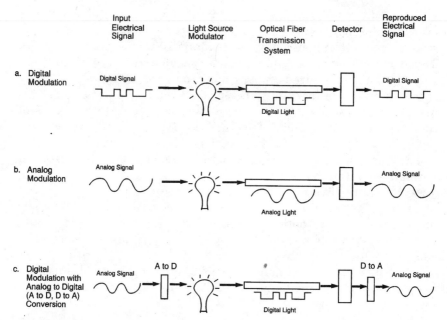

Figure 1.1 Fiber optic transmission basics.

Analog signals can also be converted to a digital format using an analog-to-digital converter (A-to-D converter) before the modulation step. Digital light signals are then propagated in the optical fiber (see Fig. 1.1c). At the other end, the digital light signal is converted to an electrical digital signal by the detector. Then a second analog-to-digital converter converts the digital signal back into its original analog form. This technique provides signal conformity with other digital signals and allows a number of signals to be combined into an optical fiber aggregate using multiplexing equipment.

The transmission techniques (in Fig. 1a, b, and c) only show information transmission in one direction. However, most systems require full, simultaneous two-way communications. Therefore, a second identical set of modulation and detection devices is implemented in the opposite direction to form a fully functional two-way communication system (see Fig. 1.2).

1.3 Advantages and Disadvantages

Optical fiber has become a popular medium for many communications requirements. Its appeal can be attributed to the many advantages optical fiber has over conventional, electrical transmission methods. This lightwave transmission medium also has drawbacks, however,

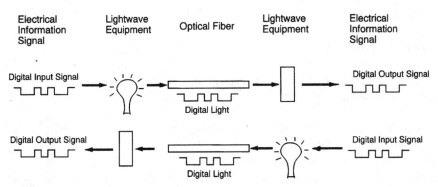

Figure 1.2 Two-way communications.

which should be examined before proceeding with an installation. The following sections describe some of these considerations.

Advantages

Large capacity. Optical fiber has the capacity to transmit large amounts of information. With present technology, 60,000 simultaneous conversations can be placed on two optical fibers. A fiber optic cable [2 cm outside diameter (OD)] can contain as many as 200 optical fibers, which would increase the link capacity to 6,000,000 conversations. In comparison to conventional wire facilities, a large multipair cable can carry 500 conversations, a coaxial cable can carry 10,000 conversations, and a microwave radio or satellite link can carry 2000 conversations.

Size and weight. Fiber optic cable is much smaller in diameter and lighter in weight than a copper cable of similar capacity. This makes it easier to install, especially in existing cable locations (such as building risers) where space is at a premium.

Electrical interference. Optical fiber is not affected by electromagnetic interference (EMI) or radio frequency interference (RFI), and it does not generate any of its own interference. It can provide a clean communication path in the most hostile EMI environment. Electrical utilities use optical fiber along high-voltage lines to provide clear communication between their switching stations. Optical fiber is also free of cross talk. Even if light is radiated by one optical fiber it cannot be recaptured by another optical fiber.

Insulation. Optical fiber is an insulator. The glass fiber eliminates the need for electric currents for the communication path. Proper all-dielectric fiber optic cable contains no electrical conductor and can provide total electrical isolation for many applications. It can eliminate interference caused by ground loop currents or potentially hazardous conditions caused by electrical discharge onto communication lines such as lightning or electrical faults. It is an intrinsically safe medium often used where electrical isolation is essential.

Security.   Optical fiber offers a high degree of security. An optical fiber cannot be tapped by conventional electrical means such as surface conduction or electromagnetic induction, and it is very difficult to tap onto optically. Light rays travel down the center of the fiber and few or none of them escape. Even if a tap is successful, it can be detected by monitoring the optical power received at the termination. Radio or satellite communication signals can easily be captured for decoding.

Reliability and maintenance.   Optical fiber is a constant medium and is not subject to fading. Properly designed fiber optic links are immune to adverse temperature and moisture conditions and can even be used for underwater cable. Optical fiber also has a long service life span, estimated at over thirty years for some cables. The maintenance required for a fiber optic system is less than for a conventional system because fewer electronic repeaters are required in a communications link; there is no copper in the cable that can corrode and cause intermittent or lost signals; and the cable is not affected by short circuits, power surges, or static electricity.

Versatility.   Fiber optic communications systems are available for most data, voice, and video communications formats. Systems are available for RS232, RS422, V.35, Ethernet, Arcnet, FDDI, T1, T2, T3, Sonet, 2/4 wire voice, E&M signal, composite video, and many more.

Expansion.   Properly designed fiber optic systems can easily be expanded. A system designed for a low data rate, for example, T1 (1.544 Mbps), can be upgraded to a higher data rate system, OC-12 (622 Mbps), by changing the electronics. The fiber optic cable facility can remain the same.

Signal regeneration.   Present technology can provide fiber optic communication beyond 70 kms (43 mi) before signal regeneration is required, which can be extended to 150 kms (93 mi) using laser amplifiers. Future technology may extend this distance to 200 kms (124 mi) and possibly 1000 kms (621 mi). The saving in intermediate repeater equipment cost as well as maintenance can be substantial. Conventional electrical cable systems, by contrast, can require repeaters every few kilometers.

Disadvantages

Electrical-to-optical conversion. Before connecting an electrical communication signal to an optical fiber, the signal must be converted to the lightwave spectrum [850, 1310, or 1550 nanometers (nm)]. This is performed by the electronics at the transmitting end, which properly format the communication signal and convert it to an optical signal using an LED or solid-state laser. This optical signal is then propagated by the optical fiber. At the receiving end of the optical fiber, the optical signal must be converted back to an electrical signal to be useful. The cost of converting the electronics should be considered in all applications.

Right of way. A physical right of way for the fiber optic cable is required. The cable can be directly buried, placed in ducts, or strung aerially along the right of way. This may require the purchase or leasing of property. Some rights of way may be impossible to acquire. For locations such as mountainous terrain or some urban environments other wireless communication methods may be more suitable.

Special installation. Because optical fiber is predominantly silica glass, special techniques are needed for the engineering and installation of the links. Conventional wire cable installation methods, for example, crimping, wire wrapping, or soldering, no longer apply. Proper fiber optic equipment is also required to test and commission the optical fibers. Technicians must be trained in the installation and commissioning of the fiber optic cable.

Repairs. Fiber optic cable that becomes damaged is not easily repaired. Repair procedures require a skilled technical crew with proper equipment. In some situations, the entire cable may need to be replaced. This problem can be further complicated if a large number of users rely on the facility. A proper system design with physically diverse routing to accommodate such contingencies is thus important.

Although there may be many advantages that favor a fiber optic installation, they should be weighed carefully against the disadvantages of each application. All the costs of the implementation and operation of fiber optic facilities should be analyzed.

1.4 Applications

Fiber optic cable is currently being used as a communication medium for many different applications. Many telephone companies, for example, are deploying fiber optics to provide communication between their central offices (COs), throughout cities, across countries, and over long oceanic routes (see Fig. 1.3). Plans now exist to extend fiber right into the home for high-quality video telephone transmissions. Fiber optics provides a reliable, high-capacity link for voice, data, and video traffic.

Cable television companies are deploying fiber optic cable to carry high-quality signals, from their head end center to hub locations distributed around cities (see Fig. 1.4). Fiber optics improves the quality of television signals and increases the number of available channels. Future plans may involve connecting optical fiber directly into the home to provide many new services for the user. Such fiber optics-based services as interactive television, banking at home, or working from a home office system are planned for the future.

Fiber optics is ideal for data communications. Very high data rates can be achieved on a thin fiber optic cable. Signals are not distorted by office interference and do not cause any interference of their own. The dielectric properties of fiber optics provide a safe interface between computers, terminals, and workstations. There is no chance of harmful ground current loops endangering users or damaging expensive computing equipment.

Many computer centers are using fiber optics to provide high-speed data communications in their LANs (see Fig. 1.5). A wide variety of

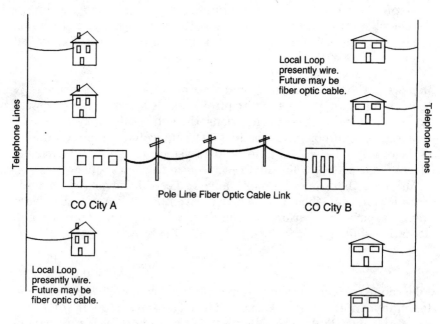

Figure 1.3 Fiber optics connecting central offices (COs).

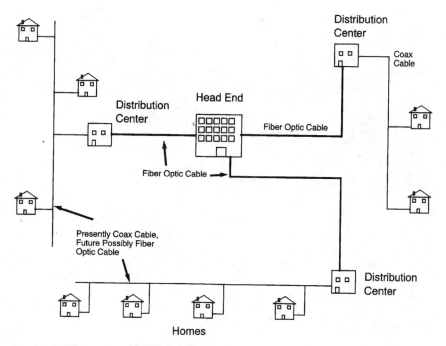

Figure 1.4   Fiber optic cable TV distribution.

products are available for many different applications. High-speed
data highways, such as fiber distributed data interface (FDDI), asynchronous transfer mode (ATM), or Sonet are available to provide
backbone connectivity with various networks. These new technologies
offer such benefits as high data transmission rates, increased distances, and reliable and secure communications. Large quantities of
data can now quickly and efficiently traverse large geographical
areas.

Because fiber optics is a cost-effective investment, companies are
installing it in metropolitan areas. These metropolitan area networks
(MANs) meet all present communications requirements and allow for
ample expansion of future systems (see Fig. 1.6).

Business can now be geographically distributed but remain connected and on line at all times. The benefits can be enormous. Banks,
for example, can keep all branches connected at high data rates for all
types of transactions. Commercial outlets can maintain centralized
control of their point-of-sale computer systems. Inventory and pricing
adjustments can be made on demand. Large firms with extensive
computing requirements, such as brokerage houses, insurance companies, and government offices, can maintain secure continuous high

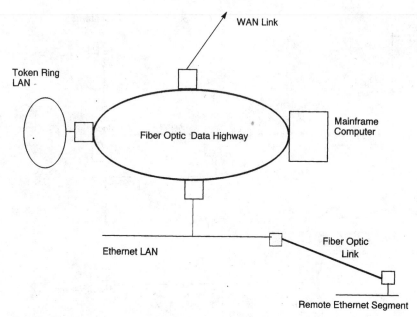

Figure 1.5 Fiber optics in a local area network (LAN) environment.

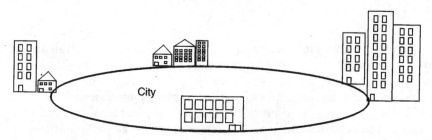

Figure 1.6 Fiber optics in a metropolitan area network (MAN).

data rate communications. Industries can keep close watch on the production and inventory of their distributed plants.

Industry is using fiber optic communication to improve the reliability and ·capacity of data and control transmissions. Because of the inherent nature of light communication it is immune to all electrical interference caused by large motors, switches, lights, and other mechanisms commonly found in industrial environments (see Fig. 1.7). Optical fiber's versatility allows computer data, telephone, video, control, and sensor transmissions all to be placed onto one fiber optic cable.

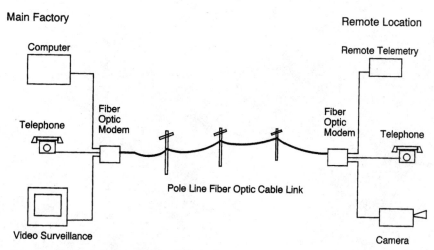

Main Factory

Remote Location

Figure 1.7   Fiber optics in an industrial environment.

2

Properties of Light

2.1 Electromagnetic Spectrum

Light behaves as an electromagnetic wave and belongs to the electromagnetic spectrum (EMS). The number of oscillations per second an electromagnetic wave completes is called its *frequency*. Visible light has a frequency of around 2.3×10^{14} cycles per second. A cycle per second is more commonly referred to as hertz (Hz). As Fig. 2.1 shows, light frequencies are much higher than other electromagnetic wave frequencies, such as radio waves and television waves.

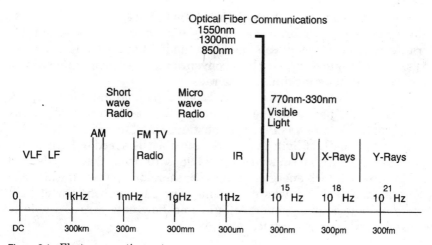

Figure 2.1 Electromagnetic spectrum.

> **Wave Theory**
>
> $\lambda = c/f$
> where λ = wavelength in meters
> c = the speed of light in the medium
> f = the frequency of light in cycles per second or hertz

The electromagnetic wavelength (λ) is the length in meters of one cycle of a wave. Visible light wavelength is in the range of 770×10^{-9} meters to 330×10^{-9} meters. One billionth of a meter, 10^{-9}, is commonly referred to as a nanometer (nm). A mathematical relationship between frequency and wavelength exists as $\lambda = c/\text{frequency}$, where c is the speed of the light in the material where it propagates. Light is normally referred to by its wavelength value in nanometers rather than its frequency.

Light used for fiber optic communication exists in the infrared (IR) region of the spectrum, just below visible light. The windows of the fiber optic communication spectrum are at 1550, 1310, and 850 nm. Light visible to the human eye begins at about 770 nm (red) and ends at 330 nm (blue). Therefore, optical fiber light is generally not visible to the eye. In some circumstances, when broad spectrum LEDs are used for 850-nm transmission, a portion of the LED's spectrum may fall into the visible range. A deep red light may then be visible.

Extreme caution must be exercised around fiber optic light. Whether visible or invisible, certain forms of fiber optic light can cause damage to the eye. Care should be exercised at all times to ensure that fiber optic light does not enter the eye directly or is indirectly reflected from a surface (see Chap. 6 on safety precautions).

The high-frequency nature of fiber optic light (2.3×10^{14} Hz) allows the light to carry information at very high data rates. Present-day transmission equipment can modulate the light at 2.4 Gbps (2.4×10^{9} bps). This is much higher than conventional electrical transmission media and is not a maximum data rate.

Light also behaves as a particle called a photon and has energy E that can be defined using the formula $E = hc/\lambda$ and expressed in units called joules. Where h is Planck's constant, equal to 6.626×10^{-34} J, λ is the wavelength of light in meters, and c is the speed of light in the propagating material (in space, $c = 2.998 \times 10^{8}$ m/s). Power is defined as the rate at which energy is delivered; therefore, power in watts can be written as $P = E/t$, where t is time. The particle behavior of light explains how sources generate light and how detectors are able to convert light back to electrical energy.

Particle Theory

$E = hc/\lambda$

where E = photon's energy in joules

h = Planck's constant, 6.626×10^{-34} J

c = the speed of light in the medium

λ = the wavelength of light in meters

2.2   Light Propagation

In free space, light travels in straight lines at a speed of 299,800 km/s or 186,292 mi/s. The direction that light waves travel is called a *light ray* and is used in fiber optics to explain many fiber characteristics.

When a light ray passes from one material to another different material, it changes speed and direction at the material's boundary. If the second material is transparent, some of the light enters the material. At the boundary between the two materials, the light ray bends before continuing in the second material. This bending is called *refraction*.

For example, if a person is standing at the edge of a lake and sees a fish in the water, the fish's physical location is different from the location it appears to have to the observer. As a light ray travels from the fish in the water (first material) to the person standing in the air above the surface (second material) it encounters the water-air boundary. At this boundary, the light ray is refracted (bent) before continuing in a straight line to the person (see Fig. 2.2).

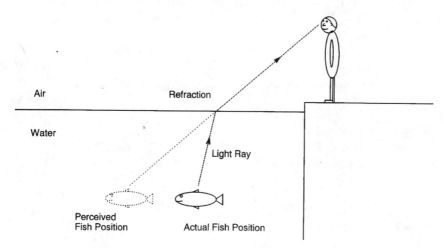

Figure 2.2 Refraction.

Because of the light ray's refraction, the fish's physical location is different from the position it appears to occupy. If the person wanted to catch the fish, he or she would therefore need to aim slightly off of the fish's perceived position. In addition, when light passes from one material to another different material, some of the light does not enter the second material but is *reflected* back into the first material.

For example, a person watching a sunset over a lake can see the sun's reflection in the lake. Light rays emitted by the sun travel directly to the person, while other sunlight rays strike the water's surface. Light rays that strike the water's surface are either reflected back into the atmosphere and seen by the observing person or refracted into the water (and seen by the fish). Light is reflected at the water-air boundary, but not completely (see Fig. 2.3).

This same behavior of light would occur if the light source were under water. Some of the light striking the water-air boundary would be reflected back into the water, and the rest would be refracted into the air. An additional phenomenon occurs in this situation. At the proper angle, all the light rays that strike the water-air boundary are reflected back into the water—none escape into the air (see Fig. 2.4). This phenomenon, also known as total reflection, is the basis for the confinement of light in an optical fiber.

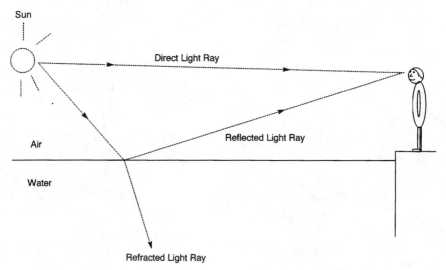

Figure 2.3 Reflection.

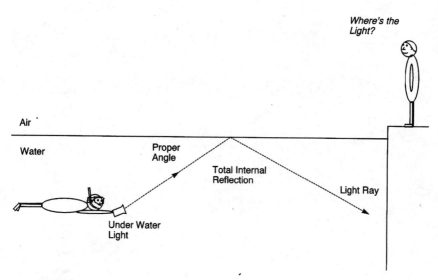

Figure 2.4 Total reflection.

3

Optical Fiber

3.1 Optical Fiber Composition

An *optical fiber* is a long, cylindrical, transparent material that confines and propagates light waves (see Fig. 3.1). It is comprised of
three layers: the center core that carries the light, the cladding layer
covering the core that confines the light to the core, and the coating
that provides protection for the cladding. The core and cladding are
commonly made from silica glass, while the coating is a plastic or
acrylate cover.

The silica core and cladding layers differ slightly in their composition due to small quantities of material such as boron or germanium
added during the manufacturing process. This alters the *index of*

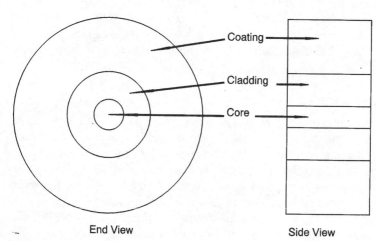

| End View | Side View |

Figure 3.1 Optical fiber.

refraction characteristic of both layers, resulting in the light confinement properties needed to propagate the light rays.

The index of refraction in the silica core is about 1.5 and the cladding is slightly less, at about 1.48. The index of refraction of air is 1.003. The fiber coating is normally colored using manufacturer's standard color codes to facilitate the identification of fiber. Optical fibers can also be made completely from plastic or other materials. They are usually less expensive but have higher attenuation (loss) and limited application.

Common fiber diameters

Optical fibers used for telecommunications are manufactured in five major core and cladding diameters, as illustrated in the following table:

Common Optical Fiber and Buffer Diameters (μm)

	Core	Cladding	Coating	Buffer or tube
I	8 to 10	125	250 or 500	900 or 2000
II	50	125	250 or 500	900 or 2000
III	62.5	125	250 or 500	900 or 2000
IV	85	125	250 or 500	900 or 2000
V	100	140	250 or 500	900 or 2000

Note: 250 μm = 0.25 mm (millimeters)

Fiber size is specified in the format "core/cladding." Therefore, a 62.5/125 fiber means the fiber has a core diameter of 62.5 μm and cladding diameter of 125 μm.

The coating covers the cladding and can be either 250 or 500 μm in diameter. For a tight-buffered cable construction, a 900-μm-diameter plastic buffer covers the coating. For a loose tube cable construction, the fiber, with a 250-μm coating, lies loose in a 2- to 3-mm plastic tube (see Chap. 4).

I. Core: 8 to 10/125 μm

A fiber with a core size of 8 to 10/125 μm is referred to as a single mode fiber. It can propagate the highest data rate and has the lowest attenuation. It is commonly used for long distance or for high data rate transmission applications. Due to the small diameter of its core, lightwave equipment employs high-precision connectors and laser light sources. This inflates equipment prices. Single mode fiber optic equipment is often priced much higher than multimode equipment. However, single mode fiber optic cable is less expensive than multimode cable.

II. Core: 50/125 μm

The 50/125-μm core size fiber was the first telecommunications fiber to sell in large quantities and is common today. Its low *numerical aperture* (NA, see Sec. 3.2) and small core size results in the least amount of source light being coupled for a multimode fiber. However, of all the multimode fibers this fiber has the highest potential bandwidth.

III. Core: 62.5/125 μm

The 62.5/125-μm fiber diameter is presently the most popular for multimode transmission and is becoming a standard for many applications. The fiber has a lower potential bandwidth than 50/125 but is less susceptible to microbending losses. Its higher NA and core diameter provides slightly better light-coupling power than the 50/125 fiber.

IV. Core: 85/125 μm

The 85/125-μm fiber is a European fiber size and is not popular in North America. It has good light-coupling ability, similar to the 100-μm core diameter, and uses the standard 125-μm diameter cladding. This allows the use of standard 125-μm connectors and splices with this fiber.

V. Core: 100/140 μm

The 100/140-μm multimode fiber's large core diameter makes it the easiest fiber to connect. It is less sensitive to connector tolerances and the accumulation of dirt on connectors. It couples the most light from the source but has a significantly lower potential bandwidth than other smaller core sizes. It can be found in intermediate-length, connector-intensive spans (in buildings) that have low data rate requirements. It is not common and may be difficult to obtain.

There are other larger core diameters, but they are less common and their applications are limited. They are used primarily for short connection spans (between equipment) or in applications other than data communications such as visual light transmission.

A summary of these fiber core sizes and their characteristics are seen in the following table:

Optical Fiber Characteristics

	Core	NA	Loss	Bandwidth	Wavelength
I	8 to 10	Smallest	Lowest	Highest	1310 or 1550
II	50	Smaller	Lower	Higher	850 or 1310
III	62.5	Medium	Low	Medium	850 or 1310
IV	85	Large	High	Lower	850 or 1310
V	100	Largest	Higher	Lowest	850 or 1310

Matching optical fibers

When matching multimode fibers for splicing or connection, the core diameters should be the same size. A 62.5/125-μm multimode fiber should only be spliced to another multimode 62.5/125-μm fiber. Single mode fibers cannot be spliced or connected to multimode fibers. When splicing single mode fibers, the *mode field diameter* is used to match the fibers instead of the core diameter.

Multimode and single mode fibers cannot be mixed or interchanged. Equipment designed for single mode fiber can only be connected to single mode fibers. Equipment designed for multimode fiber can only be connected to multimode fibers.

Fiber optic equipment specifications should be carefully reviewed to determine the proper fiber core diameter. One fiber core diameter can be specified to be used with the lightwave equipment, or a number of different sizes can be listed for the equipment.

Because of the higher light-coupling power of larger core diameters, longer transmission distances can be achieved for some applications when larger core diameters are used (for multimode fiber only).

The following table illustrates how cable length increases with an increase in core diameter (though only for some lightwave equipment). Even though optical fiber attenuation increases with an increase in core diameter, more light is coupled into the larger core diameter, and the fiber transmission length increases. Similar charts are available for selected lightwave equipment.

Fiber size (μm)	Fiber attenuation (dB/km)	NA	Cable length (km)
50/125	4.0	0.20	0.2
50/125	3.0	0.20	0.27
50/125	2.7	0.20	0.3
62.5/125	4.0	0.29	1.3
62.5/125	3.7	0.29	1.5
100/140	5	0.29	1.5
100/140	4	0.29	1.8

Whenever you use optical fibers that have not been recommended for the equipment, always consult the lightwave equipment manufacturer.

3.2   Light Transmission in a Fiber

When a light ray travels unimpeded through a medium, such as air or glass, it travels in a straight line. However, when a light ray passes from one medium to another, it bends at the medium boundary. This bending is called *refraction*. The angle at which it is refracted is called the *angle of refraction*. The angle at which the light ray strikes the medium boundary is called the *angle of incidence.*

The angle of incidence is mathematically related to the angle of refraction according to Snell's law.

Snell's Law

$n1 \times \sin a = n2 \times \sin b$
where $n1$ = the index of refraction of the first material
$n2$ = the index of refraction of the second material
a = the angle of incidence in the first material
b = the angle of refraction in the second material

Light ray refraction occurs at the end of a fiber where it passes between air and the fiber core medium. The angle of the refraction and incidence are measured to the axis perpendicular to the air fiber boundary. For a properly cleaved fiber, this axis is the same as the fiber's axis (see Fig. 3.2). *Optical fiber cleaving* is the process of cutting an optical fiber in such a manner as to produce a smooth, flat end

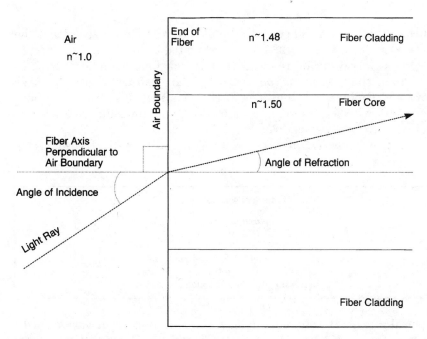

Figure 3.2   Refraction in an optical fiber.

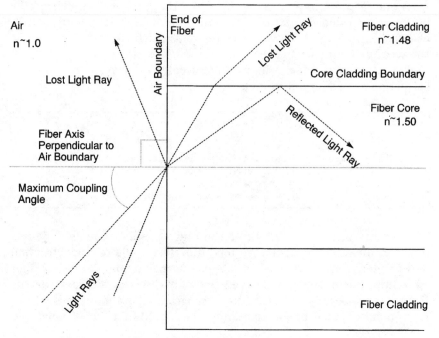

Figure 3.3   Maximum coupling angle.

surface that is perpendicular to the fiber's axis. This ensures that the maximum amount of light can be propagated in the fiber.

Only light rays that are incident on the air fiber boundary at angles less than the *maximum coupling angle* are refracted into the fiber core and captured by the fiber. Light rays incident on the air fiber boundary with angles greater than the maximum coupling angle are not captured by the fiber (see Fig. 3.3).

The fiber's *numerical aperture* (NA) is mathematically related to the maximum coupling angle.

Numerical Aperture (NA)

$$NA = \sin\,(maximum\ coupling\ angle)$$
or
$$NA = (n1^2 - n2^2)^{1/2}$$
where NA = the fiber's numerical aperture
$n1$ = the index of refraction of the core
$n2$ = the index of refraction of the cladding

Typical maximum coupling angles for a multimode fiber vary from 10 to 30°. Typical NA values vary from 0.2 to 0.5. The NA value is normally specified for an optical fiber.

When a light ray passes through a medium boundary, from a medium with a high index of refraction to a medium with a lower index of refraction, the ray is refracted at the boundary into the second medium. As the light ray's incident angle increases, a point is reached at which the light ray is no longer refracted into the second medium and is completely reflected back into the first medium. This is called *total internal reflection,* and the angle at which it occurs is called the *critical angle* (see Fig. 3.4). The critical angle can be determined by the following formula:

Critical Angle

Critical angle = arcsin ($n2/n1$)
where $n1$ = the index of refraction of the first material
$n2$ = the index of refraction of the second material

Total internal reflection only occurs when light travels through a medium boundary from a high refractive index to a lower refractive index. This phenomenon occurs at the multimode fiber's core cladding boundary and is responsible for confining the light in the fiber core.

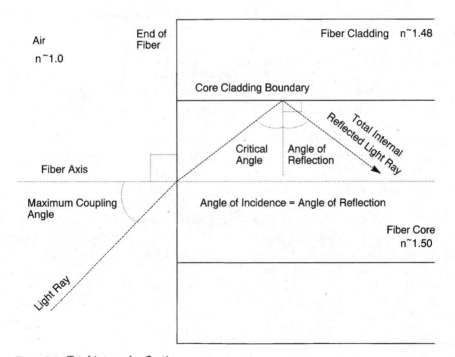

Figure 3.4 Total internal reflection.

Light rays entering the multimode fiber core at angles less than the maximum coupling angle strike the core cladding boundary at angles greater than the critical angle. They are then totally reflected back into the core and travel to the next core cladding boundary to be reflected again. The light ray's angle of reflection equals the angle of incidence at the boundary. As long as the fiber is kept straight, total internal reflection occurs and all light rays propagate in the fiber. If the fiber is bent, the angle of incidence decreases at the bend. A fraction of the light rays will decrease their incident angle past the critical angle and will be refracted into the cladding and lost.

The manufacturers of fiber specify a minimum bending radius for the fiber to make sure that the loss of the bend is negligible. This radius should be observed at all times to ensure minimum light loss and to prevent damage to the fiber.

3.3   Multimode Fiber

A multimode fiber is a fiber that propagates more than one *mode* of light. The maximum number of modes of light (light ray paths) that can exist in a fiber core can be determined mathematically by the following expression:

Number of Modes in a Fiber Core

$$M = 1 + 2D\,(n1^2 - n2^2)^{0.5}/\lambda$$

where D = core diameter
$n1$ = the core index of refraction
$n2$ = the cladding index of refraction
λ = the wavelength of light

For multimode fiber, the number of modes can easily be over 1000. The number of modes that actually exist depends on other fiber characteristics and can be reduced during propagation.

Multimode fiber is commonly used in short distance (usually less than a few kilometers) communication applications. Terminating electronics are less expensive and are typically simple in design. An LED is commonly used as the light source. Because of the multimode fiber's larger core size it is easier to connect and has greater tolerance of components with less precision.

There are two types of multimode fibers: *step index* fiber and *graded index* fiber. They differ by the index of refraction profiles of their core and cladding.

Step index fiber

A step index fiber is an optical fiber that has a core and cladding with different but uniform refractive indexes. At the core cladding boundary there is an abrupt change in the refractive index. Light confinement in any step index fiber is due to the reflection property at the core cladding boundary. This is caused by the difference in the refractive index of the two materials. As Fig. 3.5 shows, light rays are reflected at this boundary and propagated along the fiber.

The light rays travel many different paths in the fiber core. Because the distance each ray must travel differs, they therefore arrive at their destinations at different times. This results in a transmitted pulse that spreads over time.

As Fig. 3.6 shows, light rays $d1$, $d2$, and $d3$ begin at the same time t, but after traveling the length of the fiber they arrive at their destinations at different times because of the different propagation paths in the fiber core. As predicted, this results in the pulse spreading over time. This signal distortion is called *multimode (modal) dispersion*.

Such pulse-spreading restricts the data transmission rate because data rate is inversely proportional to pulse width. A wider pulse means fewer pulses can be sent per second and this results in a lower transmission bandwidth.

This is a primary data rate limiting factor in a multimode fiber:

$$\text{Data rate transmission} \propto 1/\text{Pulse width}$$

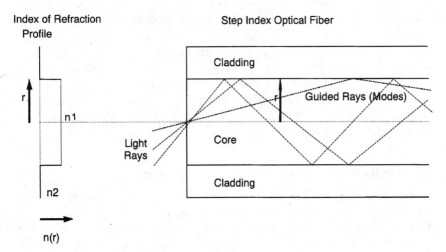

Figure 3.5 Light propagation in a step index fiber.

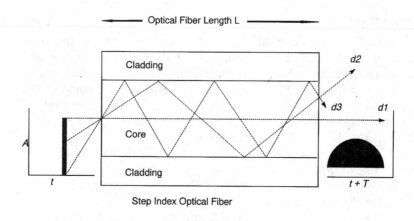

Step Index Optical Fiber

Pulse spreading in time T

Figure 3.6   Step index multimode fiber dispersion.

Graded index fiber

The refractive index of the graded index fiber core decreases from the center outward. The refractive index of the cladding is uniform. Graded index fiber bend light rays into wavelike paths because of the core's nonuniform refractive index (see Fig. 3.7). The outer region of the core has a lower refractive index than the center. Light travels faster in a material with a lower index of refraction (light velocity = c/n). Light rays traveling the longer distance, in the outer core region, require more time to get to the fiber end. However, because the light travels faster in the outer region than in the core center, the longer time caused by the distance is partially compensated for by the ray's higher speed. This reduces the amount of pulse-spreading between the core center and outer region light rays, thereby reducing the multimode dispersion. Thus, this type of fiber has a higher data rate bandwidth than the step index fiber.

3.4   Single Mode Fiber

A single mode fiber is an optical fiber that only propagates one light mode (one light ray path down the center of the fiber; see Fig. 3.8). This is accomplished by reducing the diameter of the fiber core to a size that will only permit one mode to propagate. The single mode fiber core size is between 8 and 10 μm.

The refractive index profile of single mode fiber is similar to that of the multimode step index fiber. Because of this very small core size, it is more difficult to couple light to the fiber. A solid-state laser is com-

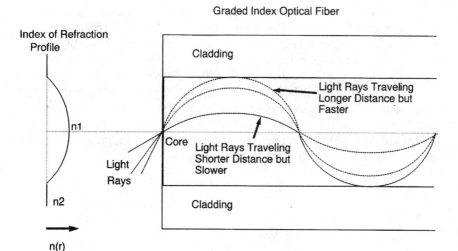

Figure 3.7 Light propagation in a graded index fiber.

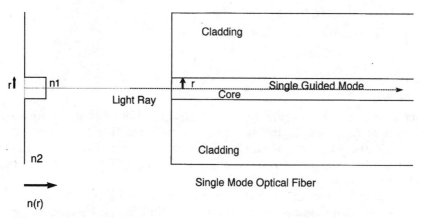

Figure 3.8 Light propagation in a single mode fiber.

monly used to accomplish this task. Higher-precision components must also be used for all fiber connections and splices.

Because only one mode is propagated in the fiber, pulse-spreading due to modal dispersion is eliminated. This enables much higher data rate transmissions over long distances. Data rates of over 2 Gbps in single mode fiber are now common. Telephone companies are one of the primary users of this technology for their long distance traffic.

3.5 Optical Power Loss (Attenuation)

Light traveling in an optical fiber loses power over distance. The loss of power depends on the wavelength of the light and on the propagating material. For silica glass, the shorter wavelengths are attenuated the most (see Fig. 3.10). The lowest loss occurs at the 1550-nm wavelength, which is commonly used for long distance transmissions.

The loss of power in light in an optical fiber is measured in *decibels* (dB). Fiber optic cable specifications express cable loss as attenuation per 1-km length as dB/km. This value is multiplied by the total length of the optical fiber in kilometers to determine the cable loss in dB.

Optical fiber light loss is caused by a number of factors that can be categorized into *extrinsic* and *intrinsic* losses:

Extrinsic

- Bending loss
- Splice and connector loss

Intrinsic

- Loss inherent to fiber
- Loss resulting from fiber fabrication
- Fresnel reflection

Bend loss

Bend loss occurs, to some degree, at all optic fiber bends due to the change of the angle of incidence at the core cladding boundary (see Sec. 3.2 and Fig. 3.9). If the bend radius is larger than the optical fiber's minimum bend radius, the loss is negligible and therefore ignored.

Bending loss can also occur on a smaller scale. Sharp curves of the fiber core, with displacements of a few millimeters or less, caused by the buffer, jacket, fabrication, installation practice, and so on, can also cause light power loss. This loss is called *microbending* and can add up to a significant amount over a distance.

Splice and connector loss

Splice loss occurs at all splices. Mechanical splices usually have the highest loss, commonly ranging from 0.2 dB to over 1.0 dB, depending on the type of splice. Fusion splices have lower losses, usually less than 0.2 dB. A loss of 0.07 dB is commonly achieved with good equipment. The loss can be attributed to a number of factors, including poor cleave, misalignment of fiber cores, an air gap, contamination, index-of-refraction mismatch, core diameter mismatch, and so on.

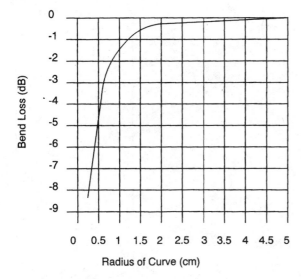

Figure 3.9 Multimode optical fiber bend loss (typical).

Losses at fiber optic connectors commonly range from 0.3 dB to over 1.5 dB and depend greatly on the type of connector used. Other factors that contribute to the connection loss include dirt or contaminants on the connector, improper connector installation, a damaged connector face, poor scribe (cleave), mismatched fiber cores, misaligned fiber cores, index-of-refraction mismatch, and so on.

Loss inherent to fiber

Light loss in a fiber that cannot be eliminated during the fabrication process is due to impurities in the glass and the absorption of light at the molecular level. Loss of light due to variations in optical density, composition, and molecular structure is called *Rayleigh scattering*. Rays of light encountering these variations and impurities are scattered in many directions and lost.

The absorption of light at the molecular level in a fiber is mainly due to contaminants in glass such as water molecules (OH^-). The ingress of OH^- molecules into an optical fiber is one of the main factors contributing to the fiber's increased attenuation in aging. Silica glass's (SiO^2) molecular resonance absorption also contributes to some light loss.

Figure 3.10 shows the net attenuation of a silica glass fiber and the three fiber operating windows at 850, 1310, and 1550 nm. For long distance transmissions, 1310- or 1550-nm windows are used. The 1550-nm window has slightly less attenuation than 1310 nm. The

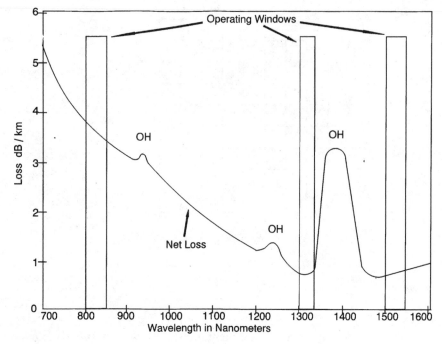

Figure 3.10 Optical fiber operating wavelengths.

850-nm communication is common in shorter distance, lower cost installations.

Loss resulting from fiber fabrication

Irregularities during the manufacturing process can result in the loss of light rays. For example, a 0.1 percent change in the core diameter can result in a 10-dB loss per kilometer. Precision tolerance must be maintained throughout the manufacturing of the fiber to minimize losses.

Fresnel reflection

Fresnel reflection occurs at any medium boundary where the refractive index changes, causing a portion of the incident light ray to be reflected back into the first medium. The fiber end is a good example of this occurrence. Light, traveling from air to the fiber core, is refracted into the core. However, some of the light, about 4 percent, is reflected back into the air. The amount being reflected can be calculated using the following formula:

Reflected Light Power at a Boundary

Reflected light (%) = $100 \times (n1 - n2)^2/(n1 + n2)^2$
where $n1$ = the core refractive index
$n2$ = the air refractive index

At a fiber connector, the amount of light power reflected is about 8 percent at the glass-air medium boundaries. This can easily be seen with an *optical time domain reflectometer (OTDR)* trace (see Fig. 13.2). The light loss generally ranges from 0.1 to 0.7 dB. This reflected light can cause problems if a laser is used and should be kept to a minimum (see Sec. 12.5).

The reflected light power can be reduced by using better connectors. Connectors with the "PC" (Physical Contact) designation are designed to minimize this reflection.

3.6   Fiber Bandwidth

Optical fiber *bandwidth* is a measure of the information-carrying capacity of an optical fiber. The optical fiber bandwidth is limited by the fiber's *total dispersion* (pulse-broadening). Dispersion limits information-carrying capacity because pulses distort and broaden, overlapping one another and become indistinguishable to the receiving equipment. To keep this from happening, pulses must be transmitted less frequently (thereby reducing the data rate).

As Fig. 3.11a shows, the original optical data pulses are discrete—ones and zeros that can be easily identified. After the signal propagates in the optical fiber for some distance (Fig. 3.11b), dispersion occurs. The pulses widen but can still be decoded by receiving equipment. Additional dispersion can introduce errors into the transmission. After further propagation in the optical fiber (Fig. 3.11c), the signal becomes totally distorted, and receiving equipment cannot derive the original waveform. Data transmission is not possible. Also, as dispersion increases the optical signal peak power is reduced, which affects the receiver's optical budget (see Sec. 12.6). Dispersion is a function of optical fiber length; the longer the fiber, the more pronounced the effect. As long as the optical fiber dispersion is within the lightwave equipment's specifications, fiber optic transmission data errors can remain below specification.

Total dispersion can be divided into two categories: chromatic dispersion and multimode dispersion (also called modal dispersion). Chromatic dispersion can be further subdivided into chromatic waveguide dispersion and chromatic material dispersion, as graphically represented here:

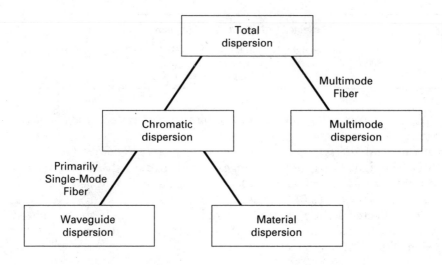

Multimode dispersion

Multimode dispersion, also known as *modal dispersion,* affects only multimode fiber and is caused by different paths, or modes, that a light ray follows in the fiber (see Fig. 3.6). This results in light rays traveling different distances and arriving at the other end of the fiber at different times. A transmitted pulse will broaden because of this effect, and this consequently reduces the maximum effective data rate.

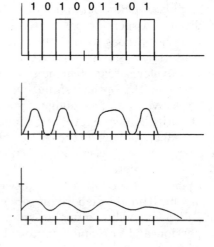

a. Digital optical signal with no dispersion. Ideal Signal.

b. Some dispersion to the same signal, but still acceptable.

c. Extensive dispersion to optical signal. Individual pulses are not distinguishable. This signal is not acceptable.

Figure 3.11 Signal dispersion.

Step index fiber has the highest modal dispersion and consequently the lowest bandwidth. Because of the nonuniform profile of index of refraction in a graded index fiber, the modal dispersion is decreased. This results in a higher transmission rate than in a step index fiber.

Chromatic material dispersion

Chromatic material dispersion occurs because the index of refraction of a fiber varies with the light wavelength in a fiber. Because a light source is comprised of a spectrum of more than one wavelength, the different wavelength's light rays travel at different speeds ($v = c/n$), resulting in pulse-broadening.

Chromatic waveguide dispersion

Chromatic waveguide dispersion is due to the spectral width of the light source when the index of refraction remains constant. The reason for this is that the geometry of the fiber causes the propagation constant of each mode to change for different wavelengths of light. Waveguide dispersion is negligible, except near zero chromatic material dispersion.

Total bandwidth

Multimode fiber bandwidth is specified by the manufacturer in the normalized modal bandwidth distance product form "megahertz $\times$ kilometer" (MHz $\times$ km). This bandwidth product accounts only for pulse-spreading due to multimode (modal) dispersion. To determine the optical fiber's total bandwidth, chromatic dispersion effects must also be considered.

Total bandwidth of multimode fiber can be calculated as follows:

$$B_{Total} = (B^{-2}{}_{Modal} + B^{-2}{}_{Chromatic})^{-1/2}$$

Single mode fiber bandwidth is only limited by the fiber's chromatic dispersion, which is specified in the form, picoseconds/(nanometers $\times$ kilometers) or (ps/nm $\times$ km). Conventional single mode fibers are available with near-zero dispersion at the 1310-nm operating wavelength (thereby supporting very high bandwidths). Optical fibers with near-zero dispersion at 1550 nm are available and are known as *dispersion shifted fibers*. Optical fibers with near-zero dispersion at both 1310 and 1550 nm are also available and are known as *dispersion flattened fibers*. For the highest data rate operation, single mode fiber should have zero dispersion at the equipment's operating wavelength.

3.7 Optical Fiber Specification: An Example

The following is an example of a typical optical fiber specification sheet available from a fiber manufacturer:

Specification		Explanation
Core diameter:	62.5 μm	Multimode fiber diameter
Cladding diameter:	125 μm	Does not include coating
Coating diameter:	250 μm	Plastic-colored coating
Mode field diameter:		This value is used for single mode fibers only.
Maximum attenuation at:		
850 nm:	3.5 dB/km	Maximum loss per kilometer
1310 nm:	1.0 dB/km	Less attenuation at longer wavelengths.
Fiber optic bandwidth:		
850 nm:	160 MHz × km	Multimode optical fiber's modal bandwidth specification at 850 nm
1310 nm:	500 MHz × km	Multimode optical fiber's modal bandwidth specification at 1310 nm
Chromatic dispersion:	0.1 ns/nm × km	This factor also limits multimode fiber bandwidth.
Cutoff wavelength:		Shortest wavelength the fiber will propagate (single mode fiber only)
Fiber manufacturer:	XYZ Company	Name of optical fiber manufacturer (not the same as fiber cable manufacturer)

4

Cable Composition

Fiber optic cable is available in two basic constructions: *loose tube cable* and *tight-buffered cable.*

4.1 Loose Tube Cable

Loose tube fiber optic cable is comprised of a number of fiber tubes surrounding a central strength member, with an overall protective jacket (see Fig. 4.1). The distinguishing feature of this type of cable is

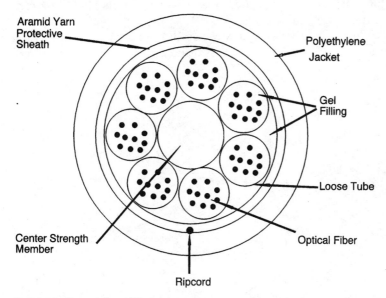

Figure 4.1 Loose tube cable.

the fiber tubes. Each tube, 2 to 3 mm in diameter, carries a number of optical fibers that lie loose in the tube. The tubes can be hollow or, more commonly, filled with a special water-resistant gel that resists water from entering into the fiber. The loose tube isolates the fiber from the exterior mechanical forces placed on the cable (see Fig 4.2).

The fibers in the tube are slightly longer than the cable itself so that the cable can elongate under tensile loads without applying stress to the fiber. During OTDR testing, this excess fiber length should be factored in to determine a more precise physical length for the cable (the excess fiber length is available from the manufacturer).

Each tube is colored, or numbered, and each individual fiber in the tube in turn is colored for easier identification. The number of fibers available in each cable range from a few to over 200.

The center of the cable contains the strength member, which can be steel, Kevlar,* or a similar material. This member provides the cable with strength and support during pulling operations as well as in permanently installed positions. It should always be securely fastened to the pulling eye during cable-pulling operations and to appropriate cable anchors in splice enclosures or patch panels.

The cable jacket can be made from polyethylene, steel armor, rubber, and aramid yarn, among other materials, for various outdoor and indoor environments. For easier and more accurate OTDR fault locating, the jacket is sequentially numbered every meter (or every foot) by the manufacturer. The pulling tension and bending radius of fiber optic cables vary, so the manufacturer's specifications should be consulted for individual cable details.

Loose tube cables are used for most outdoor installations, including aerial, duct, and direct buried applications. Loose tube cable is not

---

*Kevlar is a registered trademark of DuPont.

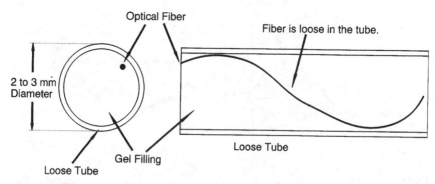

Figure 4.2   Fiber optic cable loose tube.

suitable for installation in high vertical runs because of the possibility of gel flow or fiber movement. Manufacturer's specifications should be consulted to determine the cable's maximum vertical rise for all installations. These cables are normally terminated in an appropriate fiber patch panel or splicing enclosure (see Sec. 10.6).

4.2 Tight-Buffered Cable

Tight-buffered fiber optic cable contains a number of individually buffered fibers surrounding a central strength member, with an overall protective jacket (see Fig. 4.3). The fiber buffer is a 900-μm diameter plastic cover surrounding the optical fiber's 250-μm coating (see Fig. 4.4).

The buffer provides each individual fiber with protection from the environment as well as physical support. This allows the fiber to be connectorized directly (connector installed directly onto cable fiber), without the protection offered by a splice tray. For some installations, this can decrease installation cost and reduce the number of splices in a fiber length. Because of the tight-buffer design of the cable, it is more sensitive to tensile loading and can have increased microbending loss.

Tight-buffered cable is more flexible and has a smaller bending radius than do most loose tube cables. It is designed primarily for intrabuilding installations. It can also be installed in higher vertical rises than can loose tube cable because of the individual support for each fiber. The cable is available with various fire ratings to meet

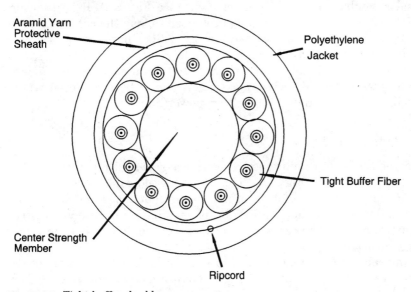

Figure 4.3 Tight-buffered cable.

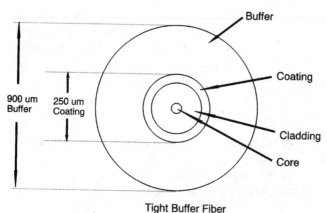

Figure 4.4   Fiber optic cable tight buffer.

standard flammability requirements. It is comparatively larger in diameter and usually more expensive than a similar loose tube cable with the same fiber count.

4.3   Figure 8 Cable

The Figure 8 cable is a loose tube cable with a factory-attached messenger. A messenger is the support member used in aerial installations. It is usually a high-tension steel cable with diameter between ¼ and ⅝ inch. The Figure 8 cable is so called because of its cross-sectional resemblance to the number eight (see Fig. 4.5). It is used for aerial installations and eliminates the need to lash the cable to a pre-installed messenger. Aerial installation of fiber optic cable is much quicker and easier with the Figure 8 cable.

The messenger is available in a high-tension steel or an all-dielectric material. The dielectric messenger should be considered when installing the cable near high-voltage power lines.

4.4   Armored Cable

Armored cables have a protective steel armor beneath a layer of polyethylene jacket (see Fig. 4.6). This provides the cable with excellent crush-resistance and rodent-protection properties. It is commonly used in direct burial applications or for installation in heavy industrial environments. The cable is usually loose tube, but a tight-buffered type is also available.

Armored cable is also available with a double-armor protective jacket for added protection in harsh environments. The steel armor

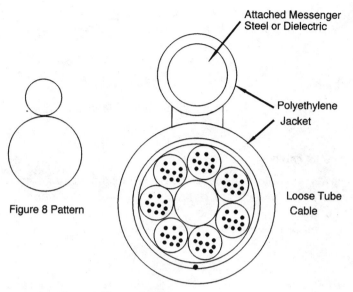

Attached Messenger
Steel or Dielectric

Polyethylene
Jacket

Loose Tube
Cable

Figure 8 Pattern

Figure 4.5   Figure 8 fiber optic cable.

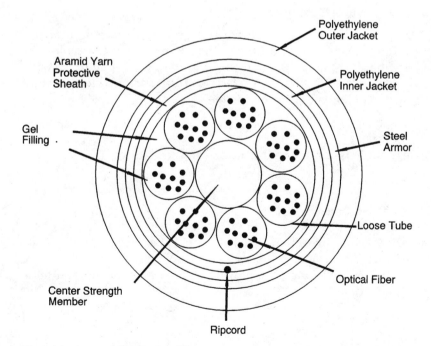

Polyethylene
Outer Jacket

Aramid Yarn
Protective
Sheath

Polyethylene
Inner Jacket

Gel
Filling

Steel
Armor

Loose Tube

Center Strength
Member

Optical Fiber

Ripcord

Figure 4.6   Armored fiber optic cable.

should always be properly grounded. At all termination points and all building entrances it should be grounded to a protective earth ground.

4.5   Other Cables

Other fiber optic cables are available for the following special applications:

Self-supporting aerial cable

Self-supporting aerial cable is a loose tube cable designed to be used in aerial installations. It does not require a messenger for support. Special clamps are used to secure the cable directly to the pole structure. The cable is placed under tension in the span.

Submarine cable

Submarine cable is a loose tube cable designed to be submerged under water. Many continents are now connected by transoceanic fiber optic submarine cable.

Optical ground wire (OPGW)

Optical ground wire (OPGW) cable is a ground wire cable that has optical fibers inserted into a tube in the center core of the cable. The optical fibers are completely protected and surrounded by heavy gauge ground wire. It is used by electric utilities to provide communications along high-voltage cable routes.

Hybrid cables

A hybrid cable is a cable that contains both optical fibers and copper pairs.

Fan-out cable

A fan-out cable is a low-fiber-count tight-buffered cable designed for easy, direct connectorization (a patch panel is not required). It is used primarily for indoor applications such as LAN connections.

The table on the facing page provides a general guide for fiber optic cable applications.

Cable Type

Application	Patch Cords	Fan Out Cable	Tight-Buffered Dielectric	Loose-Tube Dielectric	Loose-Tube Armor	Tight-Buffered Armor	Figure 8	Self-Support	Submarine
Direct connection to Equipment in same Room or Cabinet[1,2]	X	X							
Terminated by Patch Panels		X	X	X					
Between Offices Intrabuilding[2]		X[3]	X	X[3]	X	X			
Inside an Industrial Plant[2]		X	X[3]	X[3]	X	X	X	X	
High Vertical Rises		X	X			X		X	
Aerial Interbuilding				X			X	X	
Underground in Ducts				X	X[4]				
Underground Directly Buried				X[4]	X				X
Under Water				X					X
Near High Voltage			X	X					

[1] Patch cords should be placed in a dedicated cable tray or conduit when run outside of an equipment cabinet.

[2] Fire-rated cable should always be used.

[3] Cable placed in metallic conduit.

[4] Some loose-tube cable manufacturers permit cable placement in shallow water. Consult the cable manufacturer for details. Otherwise, special submarine cable is necessary.

The fiber optic cable manufacturers for each of these cable types should be contacted for further product details.

4.6 Cable Composition

Fiber optic cables are constructed with various materials to suit the installation environment. Careful consideration of cable composition is essential to prolonging cable life.

Outdoor cables should be strong, weatherproof, and ultraviolet (UV)-resistant. The cable should be able to withstand the maximum temperature variations that can be encountered during the installation process and throughout its life span. Often a cable is specified with two temperature ranges. One range specifies acceptable cable-handling and installation temperatures, and the other range indicates the cable's maximum temperature range after it has been installed and is in its final static position.

Outdoor cables are treated to inhibit UV light from penetrating the jacket and causing decomposition of the internal material. Jackets can be specified with additional UV protection if required.

Indoor cables should be strong and flexible with the required flame-retardant and smoke-inhibitor rating. Jacket colors can be bright orange or yellow for easy identification. Some of the more popular cable materials are listed in the following sections.

Polyethylene (PE)

Polyethylene is a popular cable jacket for outdoor installations. The black jacket type has good moisture- and weather-resistance properties. It is a very good insulator and has stable dielectric properties. Depending on its molecular density, it can be very hard and stiff, especially in colder temperatures. Alone it is not flame retardant, but it can be if treated with the proper chemicals.

Polyvinyl chloride (PVC)

PVC jackets offer good resistance to environmental effects, with some formulations rated for temperatures of $-55°C$ to $+55°C$. It has flame-retardant properties and can be found in outdoor as well as indoor installations. PVC is less flexible than PE and usually more expensive.

Polyurethane

Polyurethane is a common cable jacket material. Many formulations have good flame-resistance properties, and it is harder and lighter than many other materials. It also has "memory" properties, making it an ideal choice for retractable cords.

Polyfluorinated hydrocarbons (fluoropolymers)

Certain formulations of polyfluorinated hydrocarbon jacket material have good flame-resistance properties, low smoke characteristics, and good flexibility. It is used for indoor installations.

Aramid yarn/Kevlar*

Aramid yarn is a light material found just inside the cable jacket surrounding the fibers and can be used as a central strength member. The material is strong and is used to bundle and protect the individual tubes or fibers in the cable. Kevlar, a particular brand of aramid yarn, has very high strength and is often used in bulletproof vests. Fiber optic cables that must withstand high pulling tensions often use Kevlar as a central strength member. When placed just inside the cable jacket, where it surrounds the entire cable interior, it provides the fibers with additional protection from the environment. It can also provide bullet-resistance properties to the cable, which may be required in aerial cable installations running through hunting areas.

Steel armor

A steel armor jacket is often used for outdoor and indoor installations. When used in a direct burial cable, it provides excellent crush resistance and is the only material that is truly rodent-proof. For industrial environments it is used inside the plant when the cable is installed without conduits or cable tray protection. However, the steel added to the cable makes it a conductor, thereby sacrificing the dielectric advantage of the cable. Steel armor cables should always be properly grounded.

Rip cord

The cable rip cord is a strong thin thread found just below the cable jacket. It is used to easily split the cable jacket without harming the cable interior.

---

*Kevlar is a registered trademark of DuPont.

Central member

The central member is used to provide strength and support to the cable. During cable-pulling operations it should be secured to the pulling eye. For permanent installations, it should be attached to the central member anchor of the splice enclosure or patch panel.

Interstitial filling

This is a gel-like substance found in loose tube cables. It fills buffer tubes and cable interiors, making the cable impervious to water. It should be completely cleaned off with special gel-removing compound when the cable end is stripped for splicing.

4.7   Fiber Optic Cable Specification: An Example

The following is an example of a typical fiber optic cable specification sheet available from a cable manufacturer.

Specification		Explanation
Cable type:	Loose tube	
Number of fibers:	18	3 active tubes, 6 fibers per tube
Nominal weight:	166 kg/km	166 kg/km of cable
Diameter:	14.4 mm	Usually varies by 5 percent
Temperature range:		
Storage	−40 to 70°C	Storing cable on reel
Operating	−40 to 70°C	Installed operating temperature
Installation	−30 to 50°C	During installation and handling
Maximum tensile rating:		
Installation	2700 N	Maximum during installation
Permanent	600 N	Operational, no measurable change in attenuation
Minimum bend radii:		
Installation	22.5 cm	While cable is being installed
Permanent	15.0 cm	Operational, no measurable change in attenuation
Crush resistance:	220 N/cm	Operational, no measurable change in attenuation
Maximum rise:	247 m	Requires no intermediate tie-downs or loops
Copper pairs:	None	Used for communications during installation or repair
Jacket:	Polyethylene	Commonly available material
Central member:	Dielectric	Central strength member

5

Cable Procurement

Before purchasing any fiber optic cable, you should give careful consideration to all lightwave equipment and installation details so that the best cable is selected for the application. Cables are available from a standard stock, or they can be made to special order. Stock cable is usually the least expensive and has the shortest delivery time. However, the selection is usually limited to the most popular cables. Special-order cables are manufactured to the customer's specifications. They require long lead times and are more expensive.

The following list shows the information required to make a detailed order:

1.	Type of optical fiber:	Multimode or single mode
2.	Optical fiber diameter:	10/125, 50/125, 62.5/125, 85/125, 100/140 μm
3.	Operating wavelengths:	850, 1310, 1550 nm
4.	Maximum fiber attenuation:	Decibels per kilometer at operating wavelengths
5.	Minimum modal bandwidth:	Megahertz × kilometer
6.	Zero dispersion wavelength(s):	Wavelength
7.	Fiber material dispersion:	Nanoseconds/nanometers × kilometer
8.	Fiber Numerical Aperture (NA):	Value
9.	Number of optical fibers:	Number
10.	Type of fiber optic cable:	Loose tube, tight-buffered, Figure 8, patch cord, pigtail, etc.
11.	Application:	Aerial, direct burial, duct installation, etc.

12. Type of cable jacket: Outdoor, indoor, armored, etc.

13. Cable fire code rating: Plenum, nonplenum, riser, etc.

14. Cable dielectric: All-dielectric, not dielectric

15. Cable length: Meters, or feet

16. Cable price: Per-unit length, bulk discount, stock discount

17. Name of optical fiber manufacturer:

18. Name of cable manufacturer: Usually not the same as optical fiber manufacturer

19. Fiber mode field diameter: Micrometers (single mode fiber only)

20. Fiber cutoff wavelength: Nanometers (single mode fiber only)

21. Copper pairs included: Number and wire gauge

22. Cable outside diameter (OD): Millimeters

23. Cable composition: Filling compounds, strength member, armor, steel, Kevlar, etc.

24. Cable jacket: UV rating, high-voltage rating, composition

25. Operating temperature range: Maximum, minimum degrees

26. Installation temperature range: Maximum, minimum degrees

27. Minimum bend radius loaded: Centimeters

28. Minimum bend radius unloaded: Centimeters

29. Maximum vertical rise: Meters

30. Cable weight: Per-unit length

31. Maximum tension dynamic: Pounds of force

32. Maximum tension static: Pounds of force

33. Maximum cable span/sag: Meters (for aerial installations)

34. Cable markings: Sequential meter or foot marks, company name, cable identification, cable type, etc.

35. Cable crush resistance: Pounds of force/Newtons, also impact or shotgun resistance

36. Pulling eyes: Factory-installed, included in shipment

37. Cable ends: Both cable ends accessible while on the reel without unlagging the reel

38. Optical fiber color coding: Standard or special request

39. Cable shipping reel lengths: Meters, feet

40. Number of cable reels: Number

41. Type and size of cable reel: Can installation crew handle the reel size?

42. Empty reel returnable: Some reels can be returned for a
                                        refund.

43. Cable test sheet: OTDR test data available from
                                        manufacturer

44. Delivery and shipping dates: Confirmed

45. Delivery location and charges: Delivery to site

46. Cable future price: Guarantee for additional purchase

47. Other considerations: Other possible consideration in
                                        cable procurement

For simple multimode fiber applications that use standard products
and design, such as many LANs, the lightwave equipment manufacturer will specify a standard optical fiber and cable that can be used
in the application. The detail list can then be abbreviated to the following:

1. Optical fiber diameter: 50/125, 62.5/125, 85/125, 100/140
                                        µm

2. Operating wavelengths: 850 or 1310 nm

3. Maximum fiber attenuation: Decibels per kilometer at operating
                                        wavelengths

4. Minimum modal bandwidth: Megahertz × kilometers

5. Fiber Numerical Aperture (NA):   Value

6. Number of optical fibers: Number

7. Type of fiber optic cable: Loose tube, tight-buffered, Figure 8,
                                        patch cord, pigtail, etc.

8. Application: Aerial, direct burial, duct installa-
                                        tion, etc.

9. Type of cable jacket: Outdoor, indoor, armored, etc.

10. Cable fire code rating: Plenum, nonplenum, riser, etc.

11. Cable dielectric: All-dielectric, not dielectric

12. Cable length: Meters or feet

13. Cable price: Per-unit length, bulk discount, stock
                                        discount

14. Name of optical fiber manufacturer:

15. Name of cable manufacturer: Usually not the same as optical
                                        fiber manufacturer

6

Safety Precautions

There are certain precautions that should be taken when working with fiber optics. These will help maintain a safe work environment and reduce lost time due to accidents. In addition to these precautions, all other safety rules for the installation environment should also be followed. This chapter discusses the specific safety precautions that should be observed when working with fiber optics.

Cutting and stripping cable

When cutting and stripping fiber optic cable, personnel should wear proper safety glasses and gloves. Tools such as cutters, strippers, cleavers, and so on can be very sharp and can cause injury. Small pieces of cut fiber can easily fly into the air during cutting, cleaving, or scribing procedures.

Optical fiber pieces

Cut pieces of fiber from the stripping or cleaving process should be disposed of immediately in a closable container labeled "scrap fiber." Cut pieces of glass fiber are very sharp and can easily damage the eye or pierce the skin. Fibers should be handled with tweezers only.

Laser light

Fiber optic light can cause serious injury to the eye, even if it is invisible. Always turn off all light sources before working on any optical fiber. Never look into the end of a fiber that may have a laser coupled to it.

Lock out laser sources and label them "Do Not Activate" to prevent accidental activation.

Cable tension

Under tension, fiber optic cable strength members can have a lot of spring in them and can easily whip back and cause injury. Care should be taken during the cable-pulling operation or whenever the strength member is under tension.

Solvents and cleaning solutions

Liquids that are used to clean optical fibers and remove filling compound can cause skin and eye irritation. The vapors are potentially flammable and can cause respiratory problems. When working with these solvents, wear hand and eye protection, keep the area well ventilated, and don't smoke or permit open flames in the area.

Fusion splicer

The electric spark generated by a fiber optic fusion splicer can cause an explosion in the presence of flammable vapors. Never use a fusion splicer in a confined area such as an underground cable vault.

Additional safety precautions are discussed in the following chapters. Always become aware of the manufacturers' recommendations and precautions when using and installing their products.

7

Handling Fiber Optic Cable

Because of the glass properties of fiber optic cable, care should always be exercised when handling the cable. The glass fibers can easily be broken if proper handling techniques are ignored. In some instances, the cable jacket at the damaged fiber location may appear perfectly normal, making it difficult or impossible to locate without expensive instruments (an OTDR, for example, is required).

Two of the most important factors to keep in mind during an installation are the cable's *minimum bend radius* and *pulling tension.*

Minimum bend radius

Fiber optic cable has a minimum bending radius, specified by the manufacturer, for *loaded conditions,* such as during a cable pull, and for *unloaded conditions,* after the cable has been installed and is in its final resting position. The cable must not be bent tighter than the loaded minimum bending radius at any time during the installation process. The unloaded bend radius is smaller and can be used only when there is no tensile load on the cable. The bending radius varies with the cable diameter and is sometimes specified as a multiple of the cable diameter (for example, 20 × OD).

Individual fibers and fiber patch cords have a smaller minimum bending radius, usually between 3 and 7 cm (see Fig. 7.1). This minimum bending radius varies with the operating wavelength and is slightly larger for larger wavelengths (for example, 1550-nm operations).

If the fiber optic cable is bent tighter than the allowed minimum bend radius or if the cable is abused, even if no physical damage to the cable is evident the individual fibers in the cable may have been

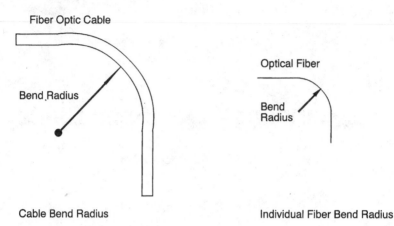

Fiber Optic Cable

Optical Fiber

Bend Radius

Bend
Radius

Cable Bend Radius   Individual Fiber Bend Radius

Figure 7.1 Bending radius.

broken or their physical characteristics altered. The abused section or
the entire length of cable may need to be replaced. The cable should
be tested immediately.

Bending the cable tighter than its minimum bend radius can damage the cable and/or increase fiber attenuation above the manufacturer's specifications.

Pulling tension

Fiber optic cable has a lower pulling tension than do many conventional cables. Maximum pulling tensions during installation are specified by the manufacturer and should not be exceeded at any time.
The cable should be pulled by hand as much as possible. Always monitor pulling tensions when using mechanical pulling techniques. A
strip-chart recorder is often used for this purpose. The cable should be
pulled in a steady, continuous motion, never jerked. Do not push the
cable at any time. The cable should be installed using the minimum
possible tension.

For permanent installed cable conditions, the tensile load on the
cable should be kept to a minimum, well below the manufacturer's
specification for tensile load in a permanent installation. Most tensile
load on a cable will occur in a vertical installation and is caused by
the cable weight. This load should be determined and kept below the
manufacturer's specifications.

General care

The cable should be handled with care at all times. Rough handling of
the cable, such as kinking, denting, abrading, or abusing, will likely

break or damage the glass fibers or alter their transmission characteristics.

Never twist the cable. If storing the cable, use a cable reel or, for short lengths, lay the cable flat in a Figure 8 pattern. Ensure that the Figure 8's curves are larger than the cable's minimum bending radius. To prevent crushing the cable when storing long cable lengths, support the cable crossing points in the middle of the Figure 8 pattern.

Avoid deforming the cable with cable clamps, supports, fasteners, guides, and the like. All clamps and supports should have a smooth, uniform contact surface. Do not tie wrap too tightly. Tie wrap only until the cable is snug. The cable jacket should not be deformed by the tie wrap.

Cables should be placed on flat trays or in conduits. Avoid causing pressure points on the cable. Also avoid laying heavy objects or piling cables on top of the fiber optic cable. All cable bends should be smooth, with radii larger than the cable's minimum bending radius. The cable should not contact any sharp objects that could possibly damage the cable.

Caution should be exercised when installing additional cables with existing fiber optic cables. Pulling eyes can have sharp edges that can easily cut through a cable jacket. Fiber optic cables should be placed in their own dedicated ducts or trays as much as possible.

For direct burial installations, the cable should lay flat in a trench, free of any larger stones or boulders that may deform the cable.

Avoid placing cable reels on their sides or subjecting them to shock from dropping. Do not allow vehicles to drive over a cable. During cable stripping, exercise care to ensure that fiber tubes or fiber buffers are not cut or damaged.

The fiber optic cable should not be cut under any circumstances unless approval has been gained from the proper authorities. Repairing fiber cable is a lengthy and costly process that introduces attenuation into the link. Fiber optic cable should be installed with minimum splices in order to keep fiber attenuation to a minimum.

Chapter

8

Outdoor Cable Installation

Fiber optic cable can be installed outdoors to link buildings or even cities together. The two most popular installations are overhead pole line installation and underground buried cable installation. An outdoor loose tube cable is commonly used in these applications. It is available with a standard outdoor jacket, an extrathick jacket (often called a double jacket), or an armored jacket.

8.1 Buried Cable Installation

Fiber optic cable can be buried directly underground or placed into a buried duct. Direct burial installations are common in long cross-country routes. Once proper plow equipment is set up, the installation process will proceed at a good pace. Special direct burial cables are used for these installations.

Alternatively, installing underground ducts can provide the cable with additional protection from the environment and can allow for future cable installation or removal without the need to dig. This can be beneficial in urban below-street installations. Also, a standard outdoor cable without armor can be used instead of the heavy armored cable.

Before digging operations begin, soil conditions along the cable route should be investigated to determine the selection of cable-placing equipment, the type of cable and duct (if used), and the installation depth. All existing underground utilities, such as buried cables, pipes, and other structures along the route should be identified and located.

The fiber optic cable can be trenched or directly plowed under. Trenching is more time-consuming than direct plowing but allows for

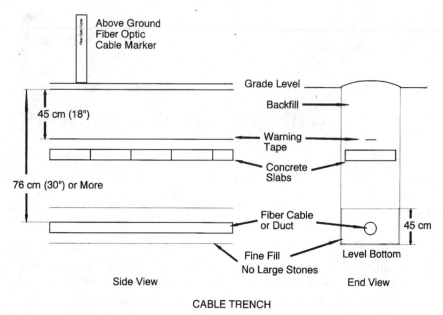

Figure 8.1 Buried fiber optic cable.

a more controlled installation. Trenches are dug by hand or by machine to depths (as shown in Fig. 8.1).

Plowing in a cable is faster but should be closely monitored to ensure that the cable is not damaged or placed near rocks during the operation. Plowing also requires that proper cable-plowing equipment is available. A combination of the two methods—plowing in isolated areas and trenching under roads or in populated areas—can also be used.

Fiber optic cable can be placed in the depth range of 30 to 40 in, depending on soil conditions, surface usage, and frost conditions (see Fig. 8.1). For example, a deeper installation of 40 in or more may be required in a farmer's field or at a road crossing. In colder climates, the cable can be buried below the frost line to avoid ground frost heaves.

Trenches should be kept as straight as possible. The bottom of the trench should be flat and level with no stones. A light backfill, with no stones, can be placed around the cable. This will provide better cable load distribution, reduce possible cable damage, and decrease optical fiber microbending loss. Backfill can be slightly above original grade level to allow for settling of the trench. All open trenches should be guarded by proper fencing.

Bright warning tape can be buried directly above the cable to alert future digging operations. Small concrete slabs (patio slabs) can also be placed above the cable in heavy traffic areas to alert backhoe operators of the buried cable. Special buried detectable tape or markers can be used to help locate buried cable. Aboveground cable markers are also available, but these may attract undesired attention to the fiber optic cable.

For direct cable burial applications, a cable with a heavy armor jacket should be selected to provide crush resistance and protection from rodents. Whenever a conducting armor is used, such as steel or aluminum, the cable should be properly grounded at all termination points and building entrances.

Splices can be stored in a watertight splice enclosure designed for direct burial installations. The enclosure is buried flush with the grade level to allow for easy entry. More elaborate systems may require full underground cable vault installation.

Minimum bending radii for cable, duct, and innerduct should always be observed.

8.2   Cable Ducts

Fiber optic cable can be pulled into new or existing ducting systems. Ducts provide the cable with protection and a means for future cable installation and removal. Ducts can also be oversized, or spare ducts can be installed to allow additional cables to be placed in the route.

When installing cable into a public duct system, the use of an innerduct will provide the cable with protection from other companies' cable installation operations (see Fig. 8.2). It also provides the cable with additional protection from the environment and can be used in old duct installations where duct "cave-ins" are common.

Most ducts and innerducts are constructed of high-density polyethylene (HDPE), PVC, or an epoxy fiberglass compound. Ducts are usually available in black or gray. Common innerduct colors are bright orange or yellow, which identify the innerducts as fiber optic.

Inside and outside walls for ducts and innerducts are available with longitudinal or corrugated ribs. These ribs help to decrease pulling tensions during installation. A smooth, outside wall is available for use in tight locations. The corrugated type is very flexible and can be used in locations with many turns. After a corrugated innerduct is pulled into a duct, it should be left for a day, uncut, to allow the innerduct to retreat back into the duct through its corrugated spring action.

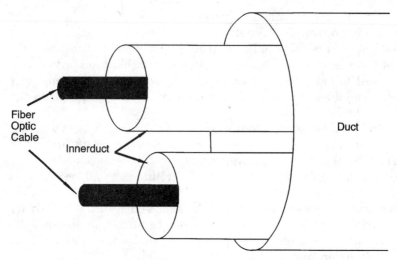

Figure 8.2 Ducts and innerducts.

Ducts and innerducts are available with pulling tape preinstalled by the manufacturer. This may save time during the installation process. They are also available prelubricated, which can dramatically decrease pulling tensions during installation.

Ducts and innerducts have a minimum bend radius. Do not bend the duct tighter than this radius. This radius can also be specified as *supported* and *unsupported*. The supported radius should be used only when the duct is bent around a supporting structure, such as in another duct or on a reel. The unsupported radius is used for duct bending in which there is no support in the bend.

After the fiber optic cable is installed in a duct or innerduct, end plugs can be installed to provide an effective water seal. The ducts and innerducts should be kept free of debris and maintained watertight at all times.

Ducts and innerducts should be sized to meet present and future cable installation requirements (see Fig. 8.3). A 50 percent fill ratio (by cross-sectional area), is a good rule of thumb to follow for a minimum duct size. For example, a 0.6-in OD cable can be installed into a 1-in inside diameter (ID) duct or innerduct (sized for single cable installation only). This size may be increased for long installation lengths with many turns. A larger duct can help reduce cable-pulling tensions. A larger duct may also be necessary to provide room for future cable installations. Standard duct sizes vary from 3 to 8 in ID. Innerduct sizes range from 0.75 to 2 in.

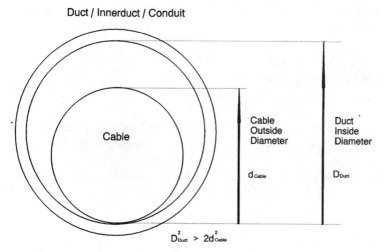

Figure 8.3 Duct/innerduct/conduit size.

Proper thick-wall duct should be used in buried applications and should be capable of withstanding the crushing force of backfill and ground traffic.

8.3 Duct Lubricant

For long cable duct pulls or pulls with numerous turns the use of a high-performance fiber optic cable lubricant is common. The primary purpose of such lubricant is to reduce the cable's *coefficient of friction* and thereby reduce the tension exerted on the cable during the pulling installation procedure.

Characteristics of fiber optic lubricant include:

- suitability for outdoor temperatures
- flame-retardant properties
- low coefficient of friction (preferably less than 0.25) when used on PE-jacketed or other types of cables
- consistent qualities over the entire installation period
- dry coefficient of friction, which can be noted for future cable pulls
- inability to affect the properties of the cable jacket, conduit, duct, or innerduct during or after installation
- tested and approved by the appropriate authorities such as UL or CSA

Lubricant should be present at all points of the duct, and it should be applied

- at all cable feed locations and intermediate pull locations
- whenever possible just before bends
- with a lubricant collar and pump

Lubricant is very slippery. All lubricant spillage should be cleaned up as soon as possible using the manufacturer's recommended procedure.

The quantity of lubricant required for an installation can be roughly estimated using the following formula:

$$Q = .00378 \times L \times (NID + NOD)$$

where Q = quantity of lubricant in liters
L = length of pull in meters
NID = nominal inner diameter of duct in centimeters
NOD = nominal outer diameter of cable in centimeters

8.4 Pulling Tape

Pulling tape is used to pull the cable or innerduct into the buried duct. The appropriate pulling tape should be used to prevent premature breakage and damage to the cable or ducting system.

Characteristics of pulling tape include:

- flat construction, similar to measuring tape, which reduces damage to duct
- each meter (or foot) marked sequentially for easy identification of distance
- Kevlar weave for greater strength
- designed to not stretch or spring
- rated for greater-than-maximum anticipated pull tension

For new duct installations, pulling tape is available preinstalled in ducts or innerducts. It can also be fished or blown into a duct length.

Pulling tape connections to cable or pulling eyes are sewn to reduce the possibility of weak links. A swivel is used between the tape and cable to prevent cable twisting. Tension-sensitive, breakable links can also be used to protect the fiber optic cable from overtension.

If additional cables will be installed in the future, pulling tape should be installed with the cable and left behind for future pulls.

8.5   Cable Installation in Ducts

Fiber optic cable can be pulled into existing or new underground ducting systems (see Fig. 8.4). Duct installation is common in populated urban centers, where constant digging in streets is discouraged, difficult, and costly. Once a ducting system is in place, cable installation can proceed with minimal interruptions.

Ducts vary in size. However, a popular diameter is 4 in. At regular intervals, ducts are terminated in underground cable vaults (manholes). These cable vaults allow the cable route to be branched in different directions and also provide a means for accessing the ducting system.

The vaults are normally rectangular and made from concrete. When a cable ducting system is created, a number of ducts are installed to provide sufficient cable capacity for present and future

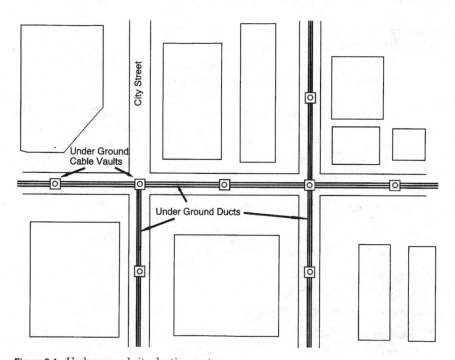

Figure 8.4   Underground city ducting system.

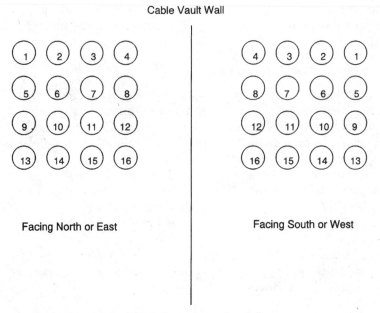

Figure 8.5 Typical duct wall configuration and assignments.

requirements. The ducts terminate on the cable vault wall in an ordered manner, as shown in Fig. 8.5.

Fiber optic cable can be pulled into an existing or new duct. The longest possible length of uninterrupted cable is pulled to reduce the number of splices. Cable lengths are determined by engineering design, considering such factors as pulling tension, route length, number of turns, pull direction, splice enclosure location, accessibility, and fiber link attenuation. Before performing any cable pull, all ducts and cable vaults should be carefully inspected for damage or deterioration and to address any safety concerns.

Before entering any cable vault, caution should be exercised to ensure a safe working environment:

- Open underground vaults should be clearly marked, guarded, and enclosed within barricades.
- Tests should be performed for dangerous gases. All vaults should be properly ventilated.
- No open flames should be permitted in or around the vault.
- Ladders, pulling irons, rails, and the like should be checked for corrosion and to ensure that they are securely fastened.
- Vehicles should be parked so that their exhaust gas does not enter the vault.

- Care should be taken so that no existing cables are damaged.

- When pumping water out of an underground vault caution should be exercised. If gasoline or oil is found, the fire department should be contacted.

- Traffic conditions should be assessed prior to entering the underground vault, and the proper permits should be obtained from city officials.

- Because the electric spark generated by a fusion splicer can cause an explosion when flammable gases are present, fusion splicers should therefore not be used in vaults.

- If vaults contain live high-voltage lines, the local utility or the company that owns the vault should be called to determine their location and to assess the dangers.

- Safety and Health regulations for industrial establishments and construction projects should always be observed.

Cable pull procedure

Perform all prescribed safety procedures and proceed as follows:

1. Open all underground cable vaults and ensure that they are safe and clear.

2. Identify all ducts to be used for the cable placement.

3. Make sure that all ducts are clear. Duct cleaning may be required.

4. If cables are present in ducts through which the fiber optic cable is to be pulled, the existing cable type should be identified and the owner of the cable called to inform him or her of the action and to identify any safety concerns.

5. To minimize cable tensions, reel vault locations should be set near the sharpest bend locations. Pulling and reel locations should also be set at corner vaults where possible. See Fig. 8.6.

6. When a cable is pulled around a bend, remember that there is added pressure against the duct wall and crushing pressure on the cable.

7. Identify the pulling and reel underground vaults. Set up all equipment appropriately (see example in Fig. 8.7).

    - The cable-pulling winch should be equipped with a strip-chart tension recorder and dynamic electronic display; winch speed should be variably controlled

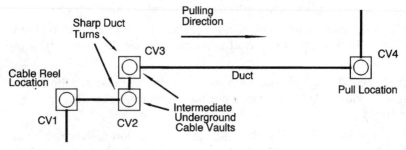

Figure 8.6 Reel and pull location.

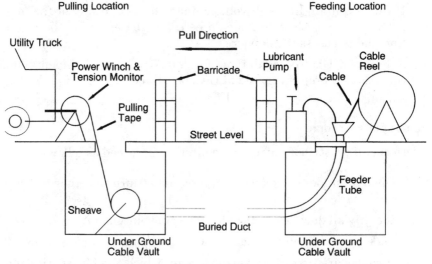

Figure 8.7 Underground duct cable pull setup.

- A lubricating collar and feeder tube should be installed in the feeding vault
- A proper-sized sheave should be installed in the pulling vault

8. A continuous length of pull tape should be installed in the duct route.

9. Before pulling cable and while the cable is still on the reel, all optical fibers in the cable should be tested with an OTDR and bare fiber adapter to ensure that they are acceptable.

10. If cable is to be replaced in innerducts, install the innerducts first:

- Position the reel of innerduct so that no obstructions prevent it from being pulled into the vault.

- Attach the proper pulling eye, with swivel, to the innerduct and attach the pulling tape to the eye.

- Ensure that all sheaves and capstans have the proper radius for the innerduct.

- During pulling operations, no personnel should be in the vaults. Extreme care should be exercised during the pull so as not to catch loose clothing, hands, or other objects in moving machinery. Loose clothing should be kept away from all moving parts. All personnel along the cable route should be in continual radio contact with each other.

- Pull as much of the innerduct by hand as possible. Apply lubricant as required. Use shorter lengths, where required, and then connect the lengths with proper innerduct connectors. Connect pulling tapes as well.

- Where hand pulls are not possible, the innerduct should be pulled with a winch. Continuous tension monitoring is not necessary, but the maximum tensions for the innerduct should not be exceeded.

- At corners or bends, sheaves may be needed to provide the innerduct with proper support. Ensure that sheaves have the proper diameter to accommodate the cable's and the innerduct's minimum bend radius.

- All lengths of innerduct and pulling tape are connected together to provide one continuous length for the cable pull.

- If a corrugated innerduct is used, do not cut it to length immediately. After the pull is complete allow the innerduct sufficient time to regain its normal shape. This will allow enough slack for the innerduct to retreat back into the duct.

- Sufficient innerduct length should be left for mounting and to allow for contraction and expansion (see the innerduct's specifications).

11. Properly attach the pulling eye and swivel to the cable. Ensure that the pulling eye and swivel assembly can easily fit through all the ducts and innerducts.

12. Do not use a woven cable grip in place of a pulling eye (except where specifications call for this; usually for short-length hand pulls).

13. Attach the installed pull tape to the swivel.

14. Adjust the capstan and sheave radius if required.

15. Pull the cable as much as possible by hand. For many cables, tension monitoring on hand pulls is not required.

16. Add lubricant generously to the cable feeder and to any mid-lubricant positions.

17. If hand pulling is too difficult, pull the cable at low speed with the winch. Avoid jerking motions. Keep pulling tensions well below the minimum cable-pulling tension. Cable tensions should be continuously monitored and recorded on a strip chart. The cable should be fed off of the reel without twisting.

18. Turn the cable reel by hand to maintain slack between the reel and the feeding collar.

19. Avoid start-and-stop actions. Greater tensions are required to start a cable from rest than to keep it in motion.

20. During the pull, if cable tension is near the maximum allowed, stop the pull and check the cable route for obstruction. Low lubricant levels or other difficulties may cause a high-tension problem.

21. After the cable tension problem is corrected, restart the pull and closely monitor the cable tension. If the tension is still high, stop the pull. The following measures can be performed to remedy the problem:

   ■ Inspect each bend closely. Ensure that they are smooth with no obstructions or sharp corners. Increase the bending radius if possible. Check to see if all sheaves turn easily.

   ■ Ensure that the feeding cable reel turns freely, with no added tension to the cable.

   ■ Try adding lubricant with a mid-lubricant pump before each bend.

   ■ Shorten the pulling route length. Move the pulling location to the route's midpoint and restart the pull. Pull the entire length of cable out through this vault and coil it on the ground. Use the Figure 8 coil pattern to reduce cable twisting. Then move the pull location back to the original location and pull the cable through the rest of the cable route.

   ■ Alternatively, use a second pulling winch at an intermediate vault location to assist in the pull. Pull out enough fiber optic cable to wrap about one to three turns around the second winch's capstan. This may introduce one to three twists into the cable.

   ■ Install sheaves in the intermediate vault as required.

   ■ By radio, coordinate both pulling winch operations.

   ■ Ensure that there are always 3 or more meters (10 or more feet) of cable slack coming off the intermediate winch before reentering the vault.

- The intermediate winch should pull the cable simultaneously with the main winch while maintaining a continuous slack loop.

- Closely monitor the cable tension at the intermediate winch.

- If maximum pull tension is recorded again, the pull location should be moved closer to the feed vault until a successful pull can be completed.

22. At intermediate vaults, when the pulling is stopped the cable should be observed to ensure that it is sufficiently lubricated.

23. Continue the pull until a sufficient length of cable (necessary for splicing outside of the vault) is pulled into the pulling vault. Additional cable may need to be pulled so that there is enough cable to be pulled back for racking in each intermediate vault.

24. Depending on engineering design, at least 6 m (20 ft) of excess cable at each end must be left coiled for future splicing and racking and for emergency repair.

25. After a sufficient length of cable has been measured at the end, the cable should be cut from the reel (ensure that proper authorization is received).

26. The cable should be tested with an OTDR to make sure it has not been damaged in the pulling process.

27. After the cable has been successfully tested, cable racking can begin. If innerduct is used, it is best to rack the innerduct in the vault without cutting it. If the innerduct is too stiff to be racked properly, then, if absolutely necessary for proper mounting, slit and remove the innerduct using the proper tools. Be extremely careful not to nick, cut, or damage the fiber optic cable inside. Rack the cable as required or clip it to the vault walls or ceiling. Place the cable as high as possible in the vault. Cable slack may need to be pulled back from the ends for racking. A corrugated slit duct can be placed around the cable for additional protection.

28. Before the innerduct is cut, ensure that an adequate amount is left over to allow for innerduct expansion. Innerduct end plugs can be used to seal the innerduct from debris and water seepage.

29. Fiber optic cable tags should be placed on the cable in each vault to identify the cable, the owner, and the owner's telephone number (see Fig. 8.8).

8.6   Aerial Installation

Aerial installation can be performed by lashing the fiber optic cable to an existing steel messenger or by installing a self-supporting fiber optic cable along a pole span (see Fig. 8.9).

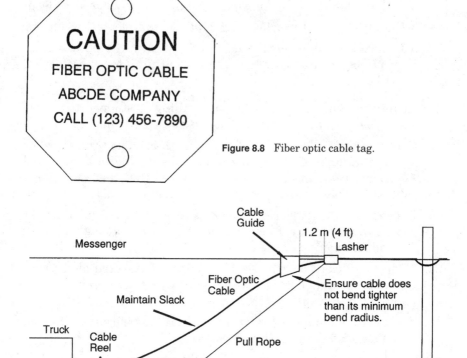

Figure 8.8   Fiber optic cable tag.

Figure 8.9   Aerial lashing setup.

Extreme caution should be observed when performing an aerial installation. The proper personnel should be contacted so that they are on the site when work is performed near high-voltage lines. Safety guidelines for aerial installation include:

- All power lines should be deenergized.
- Cables should not be installed in wet conditions.
- Cables that are installed in the vicinity of high-voltage power lines should be grounded, including all-dielectric cables.
- Proper clearance should be maintained between the fiber optic cable and power cable at all times. Always make allowances for sag due to

weather and high-voltage power line current conditions. Cable sag increases in warm weather or when power cable is passing heavy current. Overhead power lines without insulation are common.

- Make sure that all personnel are properly trained for pole line work.
- Ensure that the cable meets electric field radiation specifications.

A steel messenger is installed between the poles at the designed tension and sag to support the fiber optic cable. Avoid zigzagging the messenger from one pole side to the other. It should be kept on one side as much as possible. The messenger should always be properly grounded.

A cable reel trailer and truck is used to dispense the fiber optic cable onto the messenger. A cable guide and cable lasher are used to secure the cable to the messenger. An aerial bucket truck should follow the lasher to ensure the lasher is operating properly and to ensure that the cable is properly adjusted at pole locations. The fiber optic cable should be properly grounded before beginning any installation.

At locations where the fiber optic cable is to be placed onto a messenger with other existing cables, make sure that the lashing machine can accommodate other cables. If not, the cable can be lashed or tie wrapped manually from a bucket truck.

The cable can also be pulled onto the messenger and poles. This requires that wheel blocks be placed on each pole and along the span. The pulling rope is threaded through the blocks and attached to the cable. A winch with a tension monitor then pulls the fiber optic cable from the reel and onto the pole span. This method can be used when crossing highways, rivers, and rough terrain.

Once the cable is in place, it is lashed to the messenger. For self-supporting cable, the cable sag (tension) is adjusted to engineering specifications and is then secured to the pole and dead-end clamps.

Fiber optic cables can enter a building either by aerial or underground routes. Factors such as number of cables, appearance, bending radius, inside cable route, and security should be considered before the cables are installed.

At each pole location the cable is formed to an *expansion loop* to allow for expansion of the messenger (see Fig. 8.10). Because of the optical fiber's glass properties, fiber optic cable expands or contracts very little with changes in temperature. Therefore, to reduce tension on the fiber optic cable when it is strapped to a steel messenger, a small expansion loop is added.

The expansion loop's size is determined by engineering design (the amount required to compensate for expansion). The fiber optic cable's

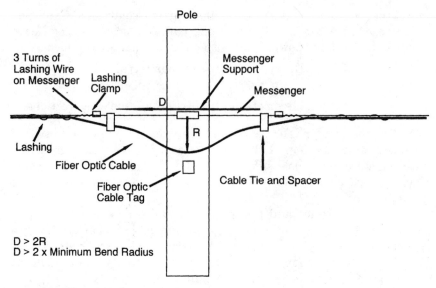

Figure 8.10 Expansion loop.

bending radius should be observed here. The length of loop D in Fig. 8.10 should be greater than twice its depth R. The length D should also be greater than twice the cable's minimum bending radius.

A bright spiral identifier cover can be placed around the expansion loop portion of the cable to identify it. A fiber optic cable tag can also be added to identify the cable.

Some new fiber optic cables available today do not require expansion loops as long as they are installed to the manufacturer's specifications. Self-supporting fiber optic cables also do not require expansion loops.

At terminating poles, where the cable is routed underground, a metallic conduit is used as a pole riser to protect the cable from damage. This conduit should be properly grounded. Depending on engineering design, at least 6 m (20 ft) of excess cable at each end is left coiled for future splicing. If splicing is required at mid-route, the weatherproof splice enclosure and the excess cable should be installed on the messenger shown in Figs. 19.2 and 19.3.

Aerial installation procedure

1. Ensure that all safety precautions are observed.

2. Install the messenger to the proper sag tension, and ensure that it is properly grounded.

3. Prepare the equipment as shown in Fig. 8.9. Install the lasher and cable guide on the messenger. The cable guide should be kept

4 ft ahead of the lasher with a stiff rod. Ensure that the cable guide's trough bend is smooth and is larger than the cable's minimum installation radius. A sheave with the proper radius can also be used.

4. Raise the cable to the cable guide and into the lasher. Keep the cable reel at least 15 m (50 ft) away from lasher. Ensure that the cable is not bent tighter than its minimum bending radius.

5. Install the lashing and secure it to the messenger with the lashing clamp.

6. To temporarily hold the cable in place on the messenger, tie-wrap the cable to the messenger at the lashing clamp.

7. Adjust the lasher for proper operation.

8. Attach a pull rope to the lasher. The lasher's pull rope should be hand pulled.

9. Begin the hand-pulling operation by pulling the lasher at a constant speed and driving the vehicle carrying the reel so that it is 50 ft from the lasher. Maintain a slight tension on the cable and keep the cable reel underneath the messenger at all times. Ensure that the proper cable bend radius is maintained at all times. Do not allow the fiber optic cable to wrap around the messenger.

10. Whenever a pole is reached, the pulling should stop. The lasher and guide are then disconnected and moved past the pole. The lashing wire is terminated with a lashing clamp, and the cable is formed into an expansion loop (if required).

11. Once the lasher and guide are set on the other side of the pole and the expansion loop is complete (as shown in Fig. 8.10), the lashing operation is continued.

12. Install fiber optic cable warning tags where required.

9

Indoor Cable Installation

9.1 Conduits and Cable Trays

Fiber optic cable can be installed in a building's existing conduit or cable tray network. In congested locations, a special dedicated conduit or cable tray may be appropriate. For situations where conduits are not practical, fire-rated armored fiber optic cable can be installed instead. The armor provides the cable with good crush resistance, and the cost of installing conduits or cable trays is eliminated. However, the armored fiber optic cable is more expensive, and the link will lose its all-dielectric properties.

Conduits and cable trays must meet mechanical restrictions imposed by the fiber optic cable. The critical constraint is the bending radius. If a cable is to be pulled into a conduit or cable tray, the conduit's bending radius must be larger than the cable's minimum bending radius for loaded conditions. If the cable is laid into a tray and will not be pulled, then the tray's bend can be as tight as the cable's unloaded minimum bending radius (see Chap. 7). All bends must have smooth curves. Cables placed in trays should be supported by a flat surface along the tray's entire length. If other cables are to be piled onto the fiber optic cable, an armored cable with high crush resistance or an added *slit duct* can be used for greater protection. A slit duct is a flexible, small-diameter duct with a longitudinal slit along its entire length. This allows the duct to be easily placed onto the cable after the cable has been put into place. This added protection will help reduce potential damage due to pressure points on the cable. All conduit and cable-tray fittings should be selected carefully to ensure that sharp edges or bends do not touch the cable at any time.

Conduits should be sized to meet present and future cable installation requirements. A 50 percent fill ratio (by cross-sectional area) is a good rule of thumb to follow for a minimal conduit diameter (see Fig. 8.3). For example, a 0.6-in OD cable can be installed into a 1-in ID conduit (sized for single cable installation only). A larger-diameter conduit may reduce pulling tensions for longer routes or for installations with many turns.

Use only conduits approved to all fire codes. When routing conduits through fire walls, ensure that the properly rated fire wall sealant is used after the installation. Conduits should also be sealed to prevent moisture, dust, or smoke migration.

A *note of caution:* Before drilling through any floors, walls, or ceilings, make sure that the drill will not contact or cut any embedded tension cables, electrical wiring, plumbing, conduits, or other objects. For masonry and other construction, x-rays of the drill area may be needed to verify the presence of a clear, unobstructed drill path.

9.2 Pull Boxes

Pull boxes are used to break up long conduit lengths for easier and lower tension pulls. Generally, they are placed at or near bends and in long straight spans. A good rule of thumb to follow is to install at least one pull box after every second 90° bend and in long conduit spans. A straight-through pull box length should be at least equal to four times the cable's minimum bend radius (see Fig. 9.1). A corner pull box length should be at least three times the cable's bending

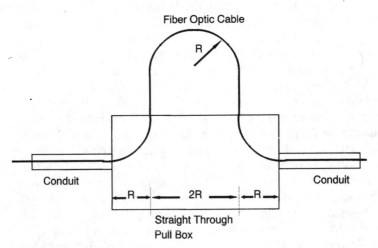

Figure 9.1 Straight-through pull box.

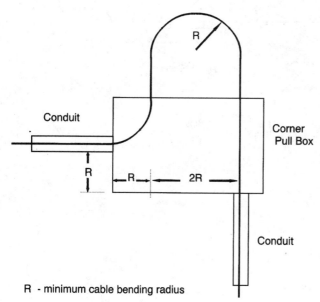

R - minimum cable bending radius

Figure 9.2 Corner pull box.

radius and one radius deep to the conduit (see Fig. 9.2). When pulling cable out of pull boxes, ensure that the cable's minimum bending radius is not compromised. Cable pulled through a corner pull box should first be pulled out into a loop (see Fig. 9.2). The sharp corner of the pull box can easily damage the cable and fibers.

9.3 Vertical Installations

In vertical installations, the weight of the cable creates a tensile load on itself. This load should not exceed the cable's maximum tensile load for a permanently installed position. A maximum cable vertical rise is also specified for a cable and should not be exceeded. Both cable load and rise conditions should be kept well below the manufacturer's specifications. Clamping a vertical cable to a support at intermediate points can reduce cable tensile loading. The clamping force should be no more than is required to prevent the cable from slipping. The cable should not deform in any way. Cable clamps should have a smooth, uniform surface to prevent cable damage.

If frequent clamping is not possible, cable hangers can be used at the top of the vertical rises and at intermediate locations along the vertical rises to suspend the cable. Cable hangers that will not damage or deform the cable should be selected. A popular choice for such

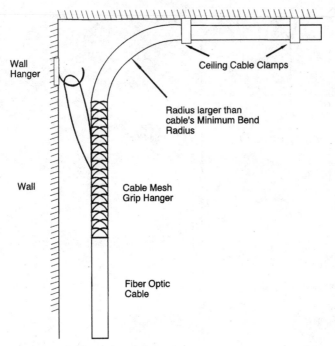

Figure 9.3 Cable mesh grip hanger.

an installation is the mesh grip or split mesh grip hanger (as shown in Fig. 9.3).

Once the mesh grip is placed onto the cable, it can be tightened to provide an effective slip-free support. It is then mounted onto a wall hanger. Make sure that the cable bend is greater than the cable's minimum bending radius at the top of the cable rise.

Tight-buffered cable is commonly used in vertical installations because of its specified high vertical rise capacity. All indoor cables should meet or exceed fire code regulations.

9.4 Building Routes

When an outdoor fiber optic cable enters a building, it should be spliced to an indoor-type fiber optic cable near the cable entrance. A splice enclosure or patch panel that can handle a number of cables for distribution can be used at this point. This provides a common optical fiber distribution point for the building and allows the proper indoor fire-rated cables to be used throughout the building. Outdoor cables can be very stiff and heavy and are difficult to install in tight building passages. It also provides a reel or pulling location for installing the

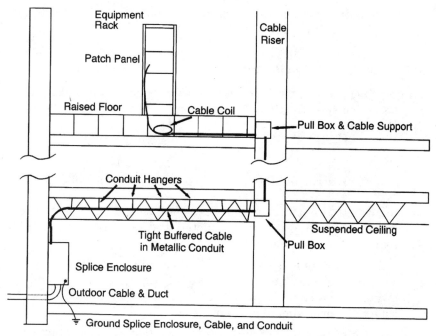

Figure 9.4 Cable route example.

outdoor cable. The splice enclosure, patch panel, conduit, and cable should all be properly grounded.

As Fig. 9.4 shows, the indoor cable runs from the splice enclosure to the equipment patch panel on an upper floor. It is normally placed into a fire-rated conduit or tray for the entire route. The conduit can be sized as shown in Fig. 8.3 or oversized to accommodate additional future cables. In some instances, the use of conduits can be prohibitive, and a fire-rated armored cable may be used instead (all dielectric cable only).

Horizontal cable routes can be placed above suspended ceilings or under raised floors. The conduit, cable tray, or cable is secured to the roof or floor with proper clamps and labeled "Fiber Optic Cable" (see Fig. 8.8).

Cable vertical rises in tall buildings are made in riser closets. When riser closets are not available, holes are drilled through the floor to route the cable or conduit. At the lightwave equipment rack, the cable can enter through the rack's top or bottom. Loops of excess fiber optic cable can be left beneath or above the rack to allow for future rack or patch panel moves or for resplicing.

Before cutting any excess fiber optic cable, proper authorization should be received. At least 6 m (20 ft) of excess cable at each end should remain for future splicing or termination. Additional length can be left uncut if required.

9.5   Cable Installation Procedure

Pulling fiber optic cable indoors is usually done by hand. Lubricant can be added in difficult pulls or when cable is placed with other existing cables.

Procedure

Perform all prescribed safety procedures and proceed as follows:

1. Identify and open all pull boxes, conduits, and cable trays and ensure that they are unobstructed and meet fiber optic cable requirements (see Secs. 9.1 and 9.2).

2. If fiber optic cable is to be placed into conduits or cable trays that contain cables, the existing cables should be identified, and the owner of the cables should be informed of the installation. All safety concerns should be identified.

3. A continuous length of pull tape is installed in the complete conduit-tray route.

4. Before proceeding with the installation, the fiber optic cable should be tested to ensure that it is acceptable (see Chap. 14).

5. Properly attach the pulling eye and swivel to the cable. Ensure that the pulling eye and swivel assembly have no sharp edges and can easily fit through all conduits, pull boxes, and cable trays.

6. Attach the installed pull tape to the swivel.

7. Hand pull the cable through the first section of conduit-tray and out of the first pull box. Make sure that cable bends are larger than the cable's minimum bending radius at all times. Add lubricant if required, avoid any jerking motions, and do not push the cable at any time.

8. Coil the cable on the floor in a Figure 8 pattern to avoid twisting. Continue the cable pull until all the cable has been pulled through the first section.

9. Feed the cable back into the pull box, and pull the cable through to the next pull box. Continue this procedure until all the cable has been installed. When feeding cable back into the pull box, make sure that the cable's minimum bending radius is not compromised.

10. If open cable trays are encountered, the cable can be laid into the tray instead of being pulled.

11. Continue this process (steps 1 to 10) until all the cable has been completely installed into the conduit-tray route.

12. Depending on the engineering design, at least 6 m (20 ft) of excess cable at each end should be left coiled for future splicing and racking.

13. Finally, the fiber optic cable should be tested to ensure that it has not been damaged during the installation (see Chap. 14).

10

Splicing and Termination

10.1 Splice Enclosures

Splice enclosures are used to protect stripped fiber optic cable and fiber optic splices from the environment. They are available for indoor as well as outdoor mounting. The outdoor type should be weatherproof, with a watertight seal.

Figure 10.1 shows a typical wall-mounted splice enclosure. The fiber optic cable is supported by cable ties, and the strength member is securely fastened to the enclosure's support. Metallic strength members are grounded.

The cable sheath stops at the splice enclosure's cable ties. Optical fiber tubes, individual thick-buffered fibers, or pigtails are supported by the tube brackets and continue on to the splicing trays. Individual optical fibers should not be exposed. The actual splices are contained in the splice trays.

10.2 Splice Trays

Splice trays (Fig. 10.2) are used to hold and protect individual fusion or mechanical splices. They are available for many types of splices, including various brand-name mechanical splices, bare fusion splices, heat-shrink fusion splices, and so on. The splice tray should be matched to the type of splice used. For example, a splice tray designed to house a mechanical splice will not hold a bare fusion splice.

Splice trays can be optical wavelength–sensitive. A splice tray designed for 810 nm may cause optical attenuation at the 1550-nm

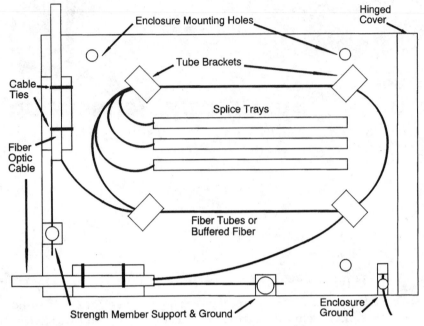

Figure 10.1 Wall-mounted splice enclosure.

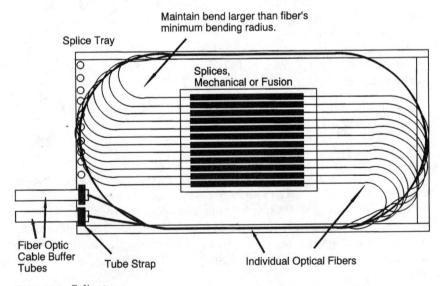

Figure 10.2 Splice tray.

wavelength. The optical operating wavelength should always be specified when ordering the trays.

Splice trays normally hold up to 12 splices, and a number of them are used together to splice a large fiber cable. Each tray terminates all the fibers in a cable buffer tube. If some fibers need to be routed to a different tray, proper tube splitters should be used. Unprotected fibers should not be exposed outside of the splice tray. Care should be exercised when mounting the splice in the tray. The individual fiber bending radius should be kept as large as possible (greater than the minimum fiber bending radius).

10.3   Patch Panels

A fiber optic *patch panel* (Fig. 10.3) terminates the fiber optic cable and allows the cable to be connected to equipment by fiber optic patch cords. It provides an access point to the equipment and the fiber cable plant. Individual fibers can be cross connected, tested, or quickly swapped between lightwave equipment. Patch panels also allow for easy fiber labeling and provide a link demarcation point.

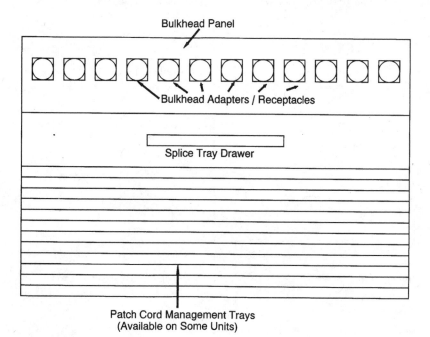

Figure 10.3 Patch panel.

The patch panel is designed with two compartments: one contains the bulkhead receptacles or adapters, and the second is used for splice tray and excess fiber storage. Patch cord management trays are optional for some patch panels and make possible the neat storage of excessive patch cord lengths.

Patch panels are available in rack-mounted or wall-mounted styles and are usually placed near terminating equipment (within patch cord reach).

If mounted in a rack, vertical location should be considered. Adequate space above or below the panel should be provided for the fiber optic cable(s) entering the enclosure. Equipment mounting, in this area, may not be possible because of the cables. The fiber optic cable's minimum bending radius should always be observed when terminating cable at patch panels.

The patch panel's *bulkhead panel* contains the *adapter* (also known as the *receptacle*). The adapter allows the cable connector to mate with the appropriate patch cord connector. It provides a low optical loss connection over many connector matings (see Fig. 10.4).

Fiber optic cable can be terminated in a patch panel using both pigtail or field-installable connector fiber termination techniques (see Fig. 10.5; see also Sec. 10.5). The pigtail technique requires that a

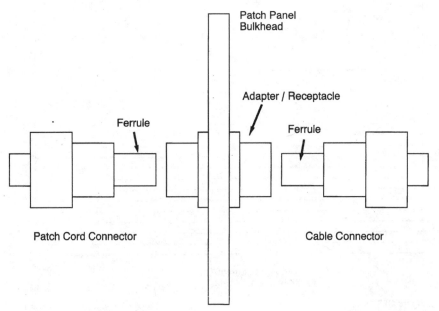

Figure 10.4 Bulkhead and adapter.

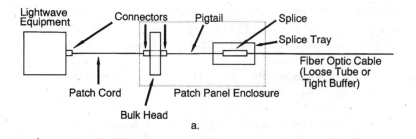

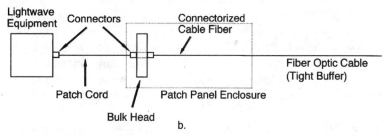

Figure 10.5   Patch panel cable termination.

splice be made and a splice tray be used. However, this technique provides for the best connection and is usually the quickest to complete.

The field-installable connector technique takes longer but does not require a splice or splice tray, thereby reducing material cost. This technique is often used with a tight-buffered cable.

Before making any fiber optic connection, ensure that all connectors and receptacles are clean. Connections should only be "finger tight." Never rotate the fiber during connection.

10.4   Splicing

Optical fiber *splicing* is the technique used to permanently join two optical fibers in a low-loss connection. This connection can be made using one of two methods: fusion splicing or mechanical splicing.

Fusion splicing provides the lowest-loss connection. The technique uses a device called a *fusion splicer* to perform the fiber splicing. The fusion splicer precisely aligns the two fibers and then generates a small electric arc to bond the two fibers together. A good fusion splicer will consistently provide low-loss splices, usually less than 0.1 dB for single mode or multimode fibers. However, such splicers are quite expensive and bulky and can be difficult to operate. For proficient operation, training by the splicer manufacturer is usually required.

A note of caution: The electric spark generated by a fusion splicer can cause an explosion when flammable gases are present. Fusion splicers should not be used in vaults or confined spaces that may contain flammable gases.

Mechanical splicing is an alternate splicing technique that does not require a fusion splicer. It uses a small mechanical splice, approximately 6-cm long and 1 cm in diameter, that permanently joins the two optical fibers. A *mechanical splice* is a small fiber connector that precisely aligns two bare fibers together and then secures them mechanically. A snap-type cover, an adhesive cover, or both is used to permanently fasten the splice. Mechanical splices are available for single mode or multimode fiber but at a higher splice loss than fusion splicing (usually less than 0.5 dB). They are small and quite easy to use and are handy for quick repairs or for permanent installations. They are available in permanent and reenterable types. Mechanical splice connection losses are greater than in fusion splicing and can range between 0.1 and 0.8 dB.

Splicing a fiber optic cable is accomplished by stripping the cable jacket with a jacket stripping tool to expose the fiber tubes or individual buffered fibers. The tube or buffer is then stripped to expose the individual coated fibers. Any gel on the tube or fibers should be carefully cleaned off. To expose the fiber cladding, the coating is stripped a short length at a time (1 cm or less). The fiber is then cleaved and spliced. The stripping process uses special hand tools that are designed not to nick or damage the individual fibers.

Cleaving a fiber provides a uniform, perpendicular surface that will allow maximum light transmission to the other fiber. Usually, high attenuation at a splice is either due to a bad splice or a bad fiber cleave. A good-quality cleaving tool is recommended.

Optical fibers can be spliced with a fusion splicer or a mechanical splice. After a splice is complete, it is stored in a splice enclosure. This protects the splice and exposed fibers and provides a means for accessing a splice if required. For small-fiber-count cables and when using thick tight-buffered fiber optic cable, connectors can be installed directly onto the cable's optical fiber, without the need for splicing and the use of splice enclosures.

Outdoor splicing of cable is commonly done in a splice van to protect the exposed fiber and equipment from the environment. It also prevents tools, equipment, and fibers from being blown around in the wind.

During cable installation, it is important to leave sufficient slack for the fiber optic cable and splice enclosure to be brought inside the van. All splicing should be done on a large, clean table with plenty of room for all splice equipment and cables. A splicing team typically consists of two technicians: a splicer and an OTDR or optical power meter operator.

Splicing tool kit

Ruler

Alcohol cleaning solution

Cable gel cleaning solution

Cotton swabs, no-residue

Paper tissues, no-residue

Fiber cleaver

Fiber coating stripper

Fiber buffer and tube stripper

Fiber jacket stripper (specify jacket size)

Utility knife

Diagonal cutters (used to cut cable or fibers to proper length)

Scissors

Emery cloth

Tweezers (used to handle cut fiber pieces)

Closable container (used to dispose of cut fiber pieces)

Protective gloves (to guard hands from cleaning solutions)

Splice protectors (for fusion splices)

Mechanical splices or fusion splicer

Splice tray and patch panel or enclosure

OTDR or power meter and light source

Large table and chairs

Splicing procedure

1. Determine the exact optical fiber assignments to be connected. Identify all optical fibers. Plan the exact fiber route in the splice enclosure, from the cable entrance into the splice tray. Ensure that the splice tray is equipped with the proper fiber holders: mechanical, bare fusion, heat shrink, crimp, and so on.

2. Strip back about 2 m (6.6 ft) of the cable jacket to expose the fiber tubes or buffered fibers (the exact length will vary for different types of splice enclosures). Use cable rip cord to cut through the

jacket. Then carefully peel back the jacket and expose the insides. If rip cord is not available, carefully strip the jacket with a jacket stripping tool or utility knife, and make sure that the optical fiber tubes or buffers are not damaged. Cut off the excess jacket. Clean off all cable gel from exposed tubes or buffers with cable gel remover. Separate the tubes and buffers by carefully cutting away any yarn or sheath. Leave enough of the strength member to properly secure the cable in the splice enclosure (see Fig. 10.6).

3. For a loose tube cable, strip away about 1 m of fiber tube using a buffer tube stripper and expose the individual fibers. The exact stripping length will vary for different splicing techniques and splice trays. For a tight-buffered cable, ensure that the individual fibers with their 900-μm buffer are exposed and loose. Be careful not to damage the optical fibers.

4. Carefully clean all fibers of any gel that may be present in the cable with the proper gel remover. Use gloves to protect your hands from the cleaning solution.

5. Identify the fiber to be spliced. Using the proper fiber coating stripper, remove enough coating so that about 5 cm of bare fiber cladding is exposed. For a tight-buffered cable remove 5 cm of buffer first with a 900-μm buffer stripping tool, and then remove the 5 cm of coating with the coating stripping tool. This length is approximate and depends on cleaver requirements and splicing method. To help hold the small fiber securely in hand while stripping, clasp the fiber with

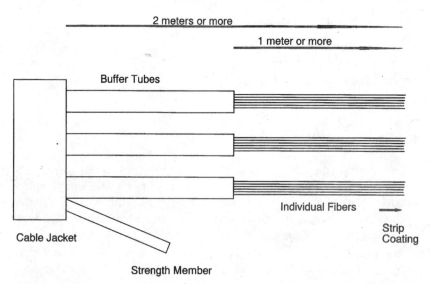

Figure 10.6 Cable stripping lengths.

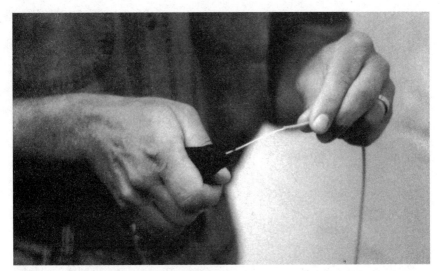

Figure 10.7 Optical fiber stripping technique.

a small piece of folded emery cloth. Always keep the coating stripping tool perpendicular to the fiber while stripping (see Fig. 10.7).

6. Carefully clean the bare fiber by wiping the fiber in one direction with a residue-free alcohol wipe (see Fig. 10.8). The use of gloves to protect your hands from the cleaning solution is recommended. After cleaning, do not touch the bare fiber with your fingers, and prevent it from touching any surface.

Figure 10.8 Optical fiber cleaning technique.

7. Prepare the cleaving tool and adjust the cleaving length as required by the splicing technique (see the manufacturer's splicing instructions).

8. Use the cleaving tool and cleave the fiber to obtain the perpendicular end surface. All fibers to be spliced must be cut with a cleaver. Using tweezers, immediately dispose of the cut fiber in a closable container designed for this purpose. (Caution: Wear safety glasses during cleaving procedure.)

9. Likewise, strip, clean, and cleave the other optical cable fiber to be spliced.

10. (a) For fusion splicing, place a splice protector, if used, on one of the fibers to be spliced. Place both fibers into the fusion splicer and follow the fusion splicer's splicing instructions. A successful splice will be mechanically strong and will have a splice loss of less than 0.1 dB as indicated by the fusion splicer, depending on the engineering design specifications. Protect the splice with the splice protector (usually a heat-shrink or crimp-on type).

10. (b) For a mechanical splice, place and align the fibers in the mechanical connector. Follow the manufacturer's splicing procedure. A successful splice will result in a splice loss that is less than that specified by the manufacturer (typically less than 0.7 dB) and will be mechanically strong. To reduce possible fiber breakage due to torsion stress, secure the fiber tube in the splice tray if possible and then loop the fiber in the splice tray to create a permanent installation before splicing.

11. After the splice is complete, carefully place the splice into the splice tray and loop excess fiber around its guides. Ensure that the fiber's minimum bending radius is not compromised and keep all bends as smooth and as large as possible.

12. At this point, an OTDR test (or power meter test) of the splice can be performed while the splicing crew is still on site. Redo the splice if required.

13. Identify the next two cable fibers to be spliced and begin the process again at step 5.

14. After all fibers have been spliced, carefully secure the fiber tubes, or buffered bundles, to the splice tray. Coil the extra length of individual fibers around the splice tray guides as shown in Fig. 10.2. Close the splice tray and place it into the splice enclosure. At all times, ensure that the fibers' minimum bend radius is not compromised.

   Wrap excess fiber tube or buffered fiber around tube brackets as shown in Fig. 10.1. Secure the fiber cable and strength member to the splice enclosure. Provide all grounding where required.

15. Test the splice with an OTDR (or power meter) from both directions. Refer to Sec. 12.4 on fiber optic facility measurement.

16. Close and mount the splice enclosure if all splices meet the engineering specifications.

10.5   Optical Fiber Termination

There are two different techniques used to terminate an optical fiber. Both are common throughout the industry. The *field-installable connector* technique is the process of terminating an optical fiber directly with a connector. Specially designed fiber optic connectors are installed directly onto the cable's fiber. The *pigtail* technique uses a factory-assembled optical fiber pigtail to terminate the fiber. Both techniques have their applications and advantages.

Field-installable connector

The field-installable connector technique allows for direct termination of optical fibers using special connectors designed for this purpose. The installation procedure involves securing the connector onto the cable's optical fiber with epoxy and then polishing the connector end to provide a low-loss connection. The end product is a cable with connectors installed directly on each fiber. The advantage of this technique is that a splice is not required for the termination. The cost of a splice and a splice tray is eliminated. The connectors are also less expensive than pigtails, but they take longer to install, thereby increasing labor cost. For an installation using a tight-buffered cable with a low fiber count, this technique can be attractive (Fig. 10.9).

The disadvantage of this technique is that it is time-consuming and not popular for single mode fiber termination. Curing of the connector glue and meticulous polishing of the fiber end are required. For multimode fiber, the resultant quality of the connection is usually good. The quality of the connection depends considerably on the technique used by the installer. Because of the very small core size of single mode fiber (10 μm) a good fiber-end polish is difficult to achieve in the field. Factory-prepared pigtails are commonly used instead.

Connectors can be installed on both tight-buffered and loose tube fiber optic cable. However, extra care should be exercised when terminating loose tube cable with connectors. Loose tube optical fibers are not protected by a buffer and can break easily if not handled properly. A fiber buffer sleeve can be slipped onto each loose tube fiber to give it additional protection and support.

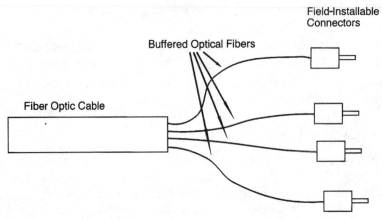

Figure 10.9 Field-installable connector.

Field-installable connector tool kit

Ruler

Alcohol cleaning solution

Cable gel cleaning solution

Cotton swabs, no-residue

Paper tissues, no-residue

Fiber scribing tool

Fiber coating stripper

Fiber buffer and tube stripper

Fiber jacket stripper (specify jacket size)

Utility knife

Epoxy

Polishing film

Polishing jig

Polishing table

Heat gun (may not be required)

Crimping tool (may not be required)

Microscope

Proper fiber optic connectors

OTDR or power meter and light source

Large table and chairs

Field-installable connector procedure

Manufacturers' installation techniques can vary for each connector type. The installer should become aware of the specific manufacturer's procedures for installing the connectors. The following describes a general procedure for a field-installable connector:

1. Determine the exact optical fiber assignments to be connected. Identify all optical fibers. Plan the exact fiber route to the equipment or patch panel. Ensure that the proper type of connectors are used.

2. Strip back about 2 m (6.6 ft) of the cable jacket to expose the fiber tubes or buffered fibers (the exact length will vary for different types of splice enclosures). Use cable rip cord to cut through the jacket. Then carefully peel back the jacket and expose the insides. If rip cord is not available, carefully strip the jacket with a jacket stripping tool or utility knife and make sure that the optical fiber tubes or buffers are not damaged. Cut off the excess jacket. Clean off all cable gel from exposed tubes or buffers with cable gel remover. Separate the tubes and buffers by carefully cutting away any yarn or sheath. Leave enough of the strength member to properly secure the cable in the splice enclosure (if required).

3. Slide the crimping collar onto the buffered fiber and/or heat shrinking tube as required.

4. Strip away about 4 cm (1.6 in) of fiber buffer as required by the manufacturer with a 900-μm buffer stripping tool.

5. Strip away about 2 cm of fiber coating as required by the manufacturer with a coating stripping tool.

6. Clean the exposed fiber with fiber cleaning solution.

7. Wipe the epoxy onto the exposed fiber as indicated by the manufacturer.

8. Slip the connector onto the fiber and push up against the buffer.

9. Slip the crimping ferrule onto the back of the connector and crimp on.

10. Place a small bead of epoxy on the front tip of the connector around the exposed fiber.

11. Allow the epoxy to cure as recommended by the manufacturer.

12. After the epoxy has cured, use the scribing tool and scribe (scratch) the fiber at the tip of the connector. (Caution: Wear safety glasses during fiber scribing.)

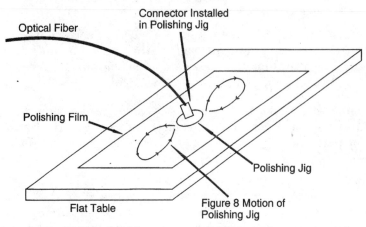

Optical Fiber

Connector Installed
in Polishing Jig

Polishing Film

Polishing Jig

Flat Table

Figure 8 Motion of
Polishing Jig

Figure 10.10   Field-installable connector polishing setup.

13. Break the fiber end away from the connector by pulling the fiber straight away.

14. Screw the connector onto the polishing jig.

15. Using the proper polishing film, place the film onto a smooth flat table. Then with light pressure polish the connector using a Figure 8 sweeping motion (see Fig. 10.10). This technique may need to be performed with two different grades of polishing film, fine and coarse. Check with the manufacturer.

16. After six or seven passes, examine the fiber end under the microscope. It should be free of scratches and glue. If scratches exist, continue polishing.

17. Place heat-shrink tubing over the back of the connector and crimped ferrule, then shrink with a heat gun.

18. Test the connector for loss using an OTDR or a light source and power meter.

Pigtail termination

The pigtail termination technique involves the splicing of a factory-assembled pigtail onto a cable fiber. This ensures a quality connector installation with low optical power loss and low return loss at the connection. Because of the required splice, the splice loss should be factored into the link budget. A splice tray and enclosure are commonly used to house the splice and connector. The factory-assembled connector provides the lowest possible optical loss, the best reliability, and the minimum return loss for both single mode and multimode fiber. The pigtail can be any length, allowing for optical fiber routing through equipment racks. This is the quickest way to terminate a fiber optic cable and can save significant time in large-fiber-count cable terminations. This method is popular for single mode fiber or loose tube cable terminations.

The drawbacks of this method are the higher cost of a pigtail compared to a field-installable connector, the need to make a splice in the fiber, and the requirement that a splice tray and patch panel or splice enclosure housing be used.

Pigtail termination tool kit

Splicing tool kit (see Sec. 10.4)

Fiber optic pigtails

OTDR or power meter and light source

Splice tray

Large table and chairs

Pigtail termination procedure (Fig. 10.11)

1. Determine the exact optical fiber assignments to be connected. Identify all optical fibers. Plan the exact fiber route to the equip-

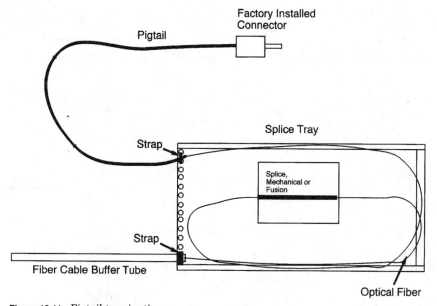

Figure 10.11 Pigtail termination.

ment. Determine fiber placement in splice tray, enclosure, or patch panel. Ensure that the proper type of connectors are used.

2. Prepare the cable end for splicing. Expose the individual fiber to be spliced. Clean, strip, and cleave the fiber as described in Sec. 10.4, on splicing procedure.

3. Prepare the pigtail for splicing. Clean, strip, and cleave each fiber pigtail as described in Sec. 10.4.

4. Splice the pigtail onto the cable fiber (see Sec. 10.4).

5. Mount splices in the splice tray and test with OTDR or power meter.

6. Install splice tray into splice enclosure or patch panel.

10.6   Fiber Optic Cable Termination

Fiber optic cable can be terminated in a number of different configurations.

Termination without enclosure

Fiber optic cable termination without an enclosure is the simplest and least costly type of termination. It is used mainly for terminating indoor tight-buffered cables with low fiber counts (normally less than six). This type of cable is light and flexible and can be run directly to terminating lightwave equipment. Each cable fiber is terminated directly with a field-installable connector (see Fig. 10.12).

The fiber optic cable end is prepared by stripping back about 1 m (3 ft) of cable jacket and other protective layers to expose the individual

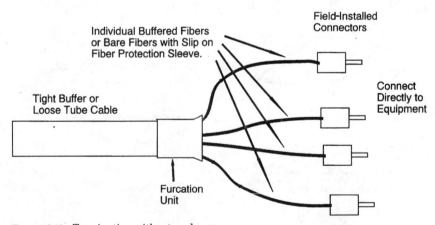

Figure 10.12   Termination without enclosure.

buffered fibers. A sleeve can be slipped onto each buffered fiber to provide it with additional support and protection. The fibers are then terminated using the field-installable technique described in Sec. 10.5. Finally, a furcation protective boot is added to the cable end to provide the fibers with some strain relief.

This cable termination technique can be used for low-fiber-count loose tube cables. However, proper fan-out termination kits should be used (available from most fiber suppliers). Loose tube cable fibers are bare, with little support, and can be broken or damaged easily. The fan-out kit includes fiber sleeves that can be slipped onto the individual fibers to provide them with protection and support. It also includes a furcation unit that provides proper loose tube and cable termination. Whenever possible, the splice tray and an enclosure should be used for loose tube termination, especially for heavy outdoor cables. Applications for this technique include dedicated LAN or video links with low fiber counts.

Termination in a splice enclosure

Termination in a splice enclosure allows for loose tube or tight-buffered cable termination using the pigtail termination technique (see Fig. 10.13). It can be used with indoor or outdoor cables with higher fiber counts. The factory-assembled pigtails have protective jackets allowing pigtail routing through cabinets or racks so the cable can be connected to lightwave equipment. Splice enclosure termination provides an effective cable termination technique using fewer components than patch panel termination (patch cords are not required) and eliminates a connection loss. However, it is not as versatile as patch panel termination.

Patch panel termination

Patch panel cable termination is the most versatile configuration. It provides quick and easy fiber identification and connection and allows patch cord connection or cross connection between equipment and other cable fibers.

The fiber optic cable can be terminated using the pigtail or field-installable connector techniques (see Sec. 10.5). Figure 10.14a diagrams the pigtail termination configuration. The cable's optical fibers are spliced to pigtails that connect to the patch panel's bulkhead adapters. Fiber optic cable can be loose tube or tight-buffered with multimode or single mode fibers.

Figure 10.14b shows the field-installable connector configuration. The cable's optical fibers are terminated in the field and then connected to the patch panel's bulkhead adapters. Splices are not required.

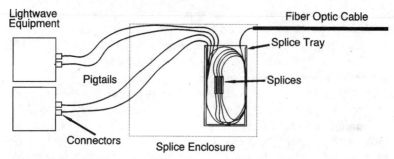

Figure 10.13 Splice enclosure termination.

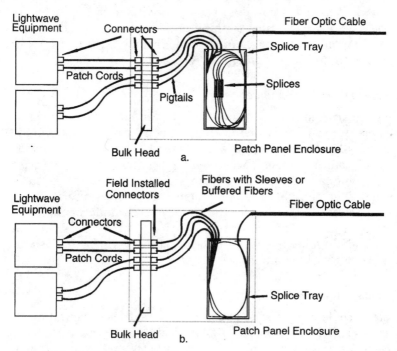

Figure 10.14 Patch panel termination.

This technique works best with tight-buffered multimode fiber. Precise field installation of connectors for single mode fiber may not be achievable. Loose tube cable can also be used in this configuration; however, the bare fibers should be protected with slip-on fiber sleeves and fanned out in a splice tray.

The following table provides a guide to the type of optical fiber jacket or buffer that can be used for various applications.

	Splice tray	Enclosure or patch panel	Cabinet, rack, or conduit	Indoors	Outdoors
Bare fiber	*				
Buffered fiber *or* fiber with sleeve	*	*			
Patch cord *or* pigtail		*	*		
Fire-rated cable			*	*	
Outdoor cable					*

11

Patch Cords and Connectors

11.1 Patch Cords and Pigtails

Fiber optic *patch cords* are analogous to electrical jumper cables. A fiber optic patch cord is short-length optical fiber with a thick, tight buffer; protective jacket; and connectors at both ends (see Fig. 11.1). The jacket is colored orange for multimode optical fiber and yellow for single mode optical fiber. It is purchased factory assembled in standard lengths or in custom lengths. The patch cords have traditionally had many uses, primarily to provide an optical connection between installed lightwave equipment and a fiber optic patch panel. Their flexibility allows for routing through tight locations in cabinets and racks full of equipment. The bending radius for a patch cord is small, usually between 2.5 and 5 cm (1 to 2 in). It can also be used for cross connecting fibers in a patch panel or for connecting test equipment to fiber optic links.

Patch cords should be laid in cable trays that are dedicated to them and not left dangling where they may be inadvertently damaged. Tie wraps can be fastened loosely around the patch cords to secure them neatly. Excess patch cord lengths can be stored in appropriate trays or tied into smooth round loops with a radius no larger than the patch cord's minimum bending radius.

If a patch cord is cut in half, each half becomes a *pigtail*. A fiber optic pigtail is used to terminate an optical fiber with a connector. The pigtail is spliced (mechanical or fusion splice) to the optical fiber to provide a factory-quality connector termination.

Patch cords should be selected to match the installed fiber optic cable's core diameter (mode field diameter for single mode fiber) and

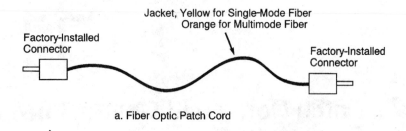

a. Fiber Optic Patch Cord

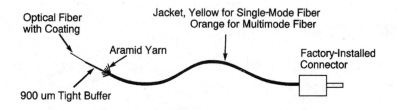

b. Fiber Optic Pigtail

Figure 11.1   Fiber optical patch cord and pigtail.

equipment connector types. If the installed fiber optic cable uses a 62.5/125 μm fiber with FC connectors, then the patch cord selected should have the same core diameter with FC connectors.

11.2   Connectors

There are a number of optical fiber connectors available today. Because lightwave equipment is not standardized on any one connector type, it is important to determine from the manufacturer the type of connector required.

A connector is comprised of a ferrule, a body, a cap, and a strain-relief boot.

The ferrule is the center portion of the connector that actually contains the optical fiber. It can be made from ceramic, steel, or plastic. For most connectors, a ceramic ferrule offers the lowest insertion loss and the best repeatability. The cap and body can either be steel or plastic. The cap can screw on, twist lock, or snap on to make a connection. The strain-relief boot relieves strains on the optical fiber.

The following list describes the various fiber optic connector types available for terminating optical fiber:

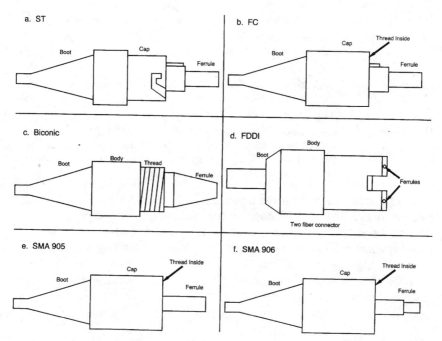

Figure 11.2   Common fiber optic connectors.

ST*   A good connector, popular for single mode and multimode
                 fiber connections, with an average loss of about 0.5 dB. It
                 has a twist locking connection that is not susceptible to loos-
                 ening in vibrating environments. It is a standard connector
                 for most fiber optic LAN equipment. See Fig. 11.2a.

FC   A good connector, popular with single mode fiber. Also known
                 as FC-PC. It has a low loss with an average of about 0.4 dB.
                 It is common in the cable TV industry. See Fig. 11.2b.

Biconic   This is an old-style connector. It was used for multimode
                 fiber but is now antiquated. It has poor repeatability and is
                 susceptible to vibrations and high loss (over 1 dB). See Fig.
                 11.2c.

SMA   This is an older connector but is still used for some equip-
                 ment. It has a higher loss of around 0.9 dB. Two types of
                 SMA connectors are available, the SMA 905 and SMA 906.
                 The SMA 905 has a straight ferrule (Fig. 11.2e). The SMA
                 906 has a stepped-down ferrule (Fig. 11.2f). The SMA 905 is
                 available with a removable collar on the ferrule, needed to
                 convert it into a SMA 906.

---

*ST is a registered trademark of AT&T, USA.

D4	This connector type is primarily used for single mode fiber.
SC	This is a new modular, high-density connector. It has low loss (under 0.5 dB) and is common in single mode installations.
FDDI	This connector is the FDDI standard fiber optic connector. It is a keyed duplex type, connecting two fibers at once. See Fig. 11.2d.
Bare fiber	This connector is used to connect an unterminated fiber. It is used to provide a temporary connection when testing bare fibers. It may require an *index matching liquid* to provide a low-loss connection.

A "PC" after the connector letter, as in FC-PC, means that the connectors make physical contact at the connection. This provides for a lower loss at the connection. Some connectors are also available that have specially treated surfaces to minimize reflected optical light. They often have the designation *super* in their names, as in Super FC-PC. They are used in single mode fiber applications with laser sources where reflected optical power can cause problems.

11.3 Cleaning Connectors

Small particles or dust can dramatically affect connector performance. A small flake of skin from the hand or scalp can easily be larger than the diameter of a single mode core. It therefore is good practice to clean a connector each time it is to be connected (see Fig. 11.3).

Cleaning items

Can of clean, compressed air with static-free nozzle

Clean, residue-free swabs

Lint-free cloth or tissue

Cleaning solvent, a residue-free alcohol solution (99 percent pure isopropyl alcohol and distilled water)

Protective gloves and safety glasses

Procedure

Safety glasses and gloves should be worn during this procedure.

1. Remove dust cap from connector receptacle or plug and wipe with a clean, alcohol-dipped cotton swab or lint-free tissue dipped in alcohol. Wipe the connector face and the inside of the sleeve.

2. Blow clean, compressed air over the entire connector surface.

3. Visually inspect the connector for cleanliness.

4. Blow the dust cap with compressed air. Place the dust cap back on the connector.

Only remove dust caps prior to making the connection. Do not let the connector touch any surface once the dustcover has been removed.

When making a connection, the connector should attach to the receptacle smoothly. Do not rotate the connector when making the connection; only rotate the threaded sleeve assembly. For screw-in connectors, tighten it only "finger tight."

When a connection is made, extremely small diameter cores are being aligned, so the procedure should be performed carefully.

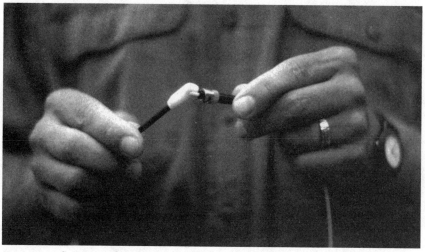

Figure 11.3 Connector cleaning.

12

Power Meter Test Procedure

The power meter test procedure is used to accurately determine fiber optic link attenuation. It should be performed as part of a final facility acceptance test or whenever a measure of link attenuation is required. Because optical fiber attenuation varies with light wavelength, the test should be conducted using the same wavelength of lightwave communication equipment. If lightwave equipment operates at the 1310-nm wavelength, the power meter and light source should also be set to 1310-nm testing.

12.1 The Decibel (dB)

The power meter test is used to determine light power loss in a fiber optic link. The measured unit of light is the *milliwatt* (mW). However, a more convenient form of measurement used is called the *decibel* (dB).

The decibel is a common measurement used in the field of electronics to determine loss or gain in a system. It is the ratio of power, voltage, or current levels between two points in logarithmic form. One point is located at the beginning or input of the system to be measured, and the other point is located at the end or output of the system. The power formula for decibel gain is expressed as:

$$G_{(dB)} = 10 \times \log \text{(output power/input power)}$$

When the output power is less than the input power, the value of this equation will always be negative. In most fiber optic applications, light power output from an optical fiber will always be less than the input light power into the optical fiber. Therefore, this value will

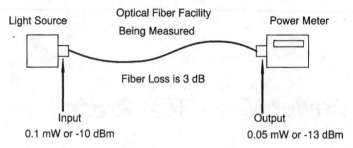

Figure 12.1   Optical power measurement.

always be negative. This negative gain can be referred to as a light loss $L_{(dB)}$.

$$L_{(dB)} = -G_{(dB)}$$

where $L_{(dB)}$ = 10 × log (input power/output power).

Light loss, $L_{(dB)}$, is a commonly used specification for fiber optic attenuation. For example, to determine the light loss of an optical fiber in a cable, a light source is connected to one end of the fiber cable (input). The light output power of the source is known to be 0.1 mW. When an optical power meter is connected to the opposite end of the fiber optic cable under test (output) the meter measures 0.05 mW (see Fig. 12.1). Using the decibel power loss formula, the optical fiber loss can be calculated as follows:

$$L_{(dB)} = 10 \times \log \text{ (input power/output power)}$$

$$L_{(dB)} = 10 \times \log \text{ (0.1 mW/0.05 mW)}$$

$$L_{(dB)} = 3 \text{ dB}$$

The light power loss of this optical fiber is 3 dB.

The dB unit is a logarithmic ratio of input and output levels and is therefore not absolute (i.e., has no units). An absolute measure of power in decibels can be made in the dBm form. The dBm unit is a logarithmic ratio of the measured power to 1 mW of reference power.

The formula is as follows:

$$P_{(dBm)} = 10 \times \log \text{ (power/1 mW)}$$

The same result in loss can be achieved using the dBm. In the previous example, light power input by the source to the optical fiber is 0.1 mW, which is −10 dBm.

$$P_{\text{Source(dBm)}} = 10 \times \log (0.1 \text{ mW/1 mW})$$

$$P_{\text{Source(dBm)}} = -10 \text{ dBm}$$

The light power received by the meter from the optical fiber's output is 0.05 mW, which is −13 dBm.

$$P_{\text{Receive(dBm)}} = 10 \times \log (0.05 \text{ mW/1 mW})$$

$$P_{\text{Receive(dBm)}} = -13 \text{ dBm}$$

The light power loss in the fiber is equal to the light source power minus the received meter light power:

$$L_{\text{(dB)}} = P_{\text{Source(dBm)}} - P_{\text{Receive(dBm)}}$$

$$L_{\text{(dB)}} = -10 \text{ dBm} - (-13 \text{ dBm})$$

$$L_{\text{(dB)}} = 3 \text{ dBm}$$

Therefore, the light power lost by the optical fiber is 3 dBm, or 3 dB.

All measurements must be in either decibels or in milliwatts but *not both*. Normally, all measurements are made in the decibel scale because it is easier to work with. It is not necessary to convert between mW and dBm because most specification data also use the decibel scale. The following table shows the figures for converting dBm into mW.

Power in dBm	Power in mW
0	1
−3	0.5
−10	0.1
−20	0.01
−30	0.001
−40	0.0001
−50	0.00001

It is helpful to remember that a loss of 3 dB is equivalent to a 50 percent loss in power. A loss of 10 dB is equivalent to a power loss of 90 percent. When you subtract two dBm values, the result is in dBm, or dB. When you add or subtract a dB to or from a dBm, the result is in dBm. When you add or subtract two dB values, the result is always in dB. Decibel values are never multiplied together—always add or subtract.

When measuring the loss in dB of a number of different sections of a fiber optic link, the total loss of all sections is equal to the sum in dB of each individual section.

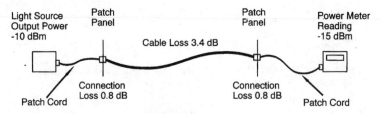

Figure 12.2 Fiber optic loss example.

Example 12.1   A fiber optic link with a 1-km cable has a loss of 3.4 dB. Patch panel connection loss at each end is 0.8 dB. Pigtail loss is negligible. If a light source with optical power of −10 dBm is connected to one end of the fiber link, what will the received light power be at the other end? (See Fig. 12.2.)

First, the total link loss including patch panel connections is summed:

$$3.4 \text{ dB} + 0.8 \text{ dB} + 0.8 \text{ dB} = 5.0 \text{ dB}$$

The optical power loss formula needs to be rearranged to equal the received optical power:

$$L_{(dB)} = P_{Source(dBm)} - P_{Receive(dBm)}$$

becomes

$$P_{Receive(dBm)} = P_{Source(dBm)} - L_{(dB)}$$
$$P_{Receive(dBm)} = -10 \text{ dBm} - 5 \text{ dB}$$
$$P_{Receive(dBm)} = -15 \text{ dBm}$$

Therefore, the light power that would be measured by an optical power meter at the end is −15 dBm. It should be noted that two fiber optic connectors contribute to one connection loss.

12.2   Equipment

For most power meter measurements, the following equipment is required:

Optical power meter:

- with proper wavelengths
- with proper connectors
- for single mode or multimode fiber sizes
- calibrated in dBm, preferably with optional dB scale

Optical light source:

- with stable light source
- with proper wavelengths

- with proper connectors
- with single mode or multimode fiber sizes
- with laser or LED source
- with sufficient output light power

Test patch cords:

- at least two, 1 to 5 m in length
- with known loss
- with proper connectors
- with same core size as fiber optic facility to be tested

Connector cleaning solution, cotton swabs, compressed air
Fiber optic cable-stripping tool kit (if required for unterminated fibers)
Bare fiber adapter (if fiber is not terminated)
Index matching gel:

- for bare fiber adapter

12.3 Patch Cord Losses

Before proceeding with optical power facility measurements, individual patch cord loss should be tested. Loss measurements compared to records or manufacturer's specifications can identify faulty patch cords. Always clean all optical connectors before testing.

Procedure

1. Turn on the power meter and light source test equipment and allow them to warm up and stabilize. If a laser light source is used, make sure that it remains off until all fibers to the source are properly terminated. Clean all connectors.

2. Switch the power meter to the dBm scale. Connect a known, good-quality test patch cord, as shown in Fig. 12.3a, to obtain a reference light source power meter reading in dBm ($P_{\text{Ref(dBm)}}$). This value should be close to the light source output power specification. Switch the power meter to the dB scale and calibrate to 0.0 dB (see the meter's instructions). Once calibrated, do not turn off or adjust the power meter.

   If the power meter does not have a dB scale and only displays the absolute power levels in dBm, record the power meter reading as $P_{\text{Ref(dBm)}}$. This is required for loss calculation (see step 6).

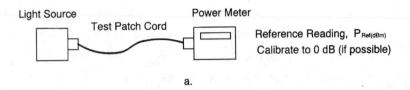

a.

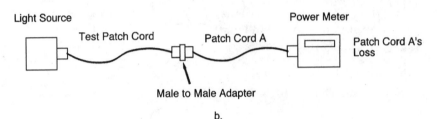

Male to Male Adapter

b.

Figure 12.3 Patch cord loss.

3. Connect the patch cord to be tested (patch cord A as shown in Fig. 12.3*b*) between the power meter and the test patch cord. A male-to-male adapter will be required for this connection.

4. Record the power meter light loss reading in dB ($L_{\text{Meter(dB)}}$). For some power meters, the light loss reading in dB may be a negative number. This indicates that the power meter uses the power formula for decibel gain instead of the light loss decibel formula (see Sec. 12.1). If the reading is negative, disregard the negative and only use the positive value for all calculations.

   If the power meter does not have the dB scale, then a quick calculation is needed to determine the patch cord loss (see step 6). Record the dBm value as $P_{\text{Meter(dBm)}}$.

5. Reverse patch cord A's connection and confirm that the reading is the same as previously. If the reading differs when it is reversed, try a different patch cord. Patch cord A's connectors may be out of alignment.

6. If the power meter reading is in dB, then this value is patch cord A's loss ($L_{\text{(dB)}}$).

   If the reading is in dBm, then subtract the received reading from the reference reading to obtain patch cord A's loss ($L_{\text{(dB)}}$):

$$L_{\text{(dB)}} = P_{\text{Ref(dBm)}} - P_{\text{Meter(dBm)}}$$

7. A good patch cord should have a loss of less than 1.0 dB. For each cord, repeat each step beginning with step 3 until all required patch cords are tested and losses are recorded (patch cords B, C, etc.).

8. The patch cord loss is the optical loss of the patch cord's fiber and both connectors. Note that the loss of two connectors adds up to a connection loss.

12.4 Fiber Optic Facility Measurement

A power meter loss measurement should be performed on all optical fiber links in order to determine the total link loss. This information is used to determine the optical link budget and optical margin (see Sec. 12.6).

Two configurations can be used for this test: the loop back or the end-to-end configuration.

The end-to-end configuration is more accurate, but it requires two people to perform. The loop back configuration provides an averaged result and can be completed with one person, but it is more time-consuming.

Ensure that the test patch cords are used. If a patch panel is not used, the lightwave equipment connects directly to the terminated fiber optic cable; then, connect the terminated fiber optic cable directly to the test equipment.

Procedure

1. Turn on the power meter and light source test equipment and allow them to warm up and stabilize. If a laser light source is used, ensure that it remains off until all fibers to the source are properly terminated (see Chap. 6 on safety precautions). Clean all connectors.

2. Identify the individual optical fibers to be tested. For an installed operating system, power down all connected communication equipment and ensure that all optical fiber light sources are off.

3. Before the facility can be tested, the power meter should be calibrated to 0 dB as follows. Using a test patch cord, connect the light source to the power meter as shown in Fig. 12.3a. Set the power meter to the dBm scale. Make sure that the light source is on, and read the received optical power at the power meter in dBm. This value should be close to the light source manufacturer's output power specification. Set the power meter to the dB scale and adjust to 0.0 dB. This 0-dB calibration will be used as the light source's reference power level. Disconnect this test assembly but do not adjust or turn off the power meter. If the power meter does not have a 0-dB calibration and only displays the absolute power

levels in dBm, then record the power meter reading $P_{\text{Ref(dBm)}}$ for later calculations.

4. For an existing system, disconnect all existing patch cords from the fiber optic cable patch panel.

   Determine test configuration (end-to-end or loop back) and connect the tested patch cords, light source, and power meter (see Fig. 12.4). Do not disturb the power meter's zero reference calibration, performed in the previous step.

   In an end-to-end test configuration, connect the light source using the test patch cord to one end of the fiber link and power meter using a second tested patch cord (patch cord A) to the other end of the fiber link (see Fig. 12.4b).

   If a loop back test configuration is used, add loop back patch cord B to patch panel B and set up as shown in Fig. 12.4a. Radio or telephone communication can be established between the two testing locations to speed up the procedure.

5. Ensure that the configuration is connected properly, and turn on the light source. Read the optical power meter and record the optical power level. For the end-to-end configuration, record the loss in dB as $L_{\text{MeterEnd(dB)}}$. If the dB scale is not used, record the power level in dBm as $P_{\text{MeterEnd(dBm)}}$. If the loop back configuration is used,

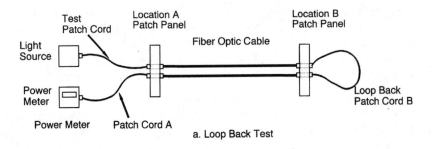

a. Loop Back Test

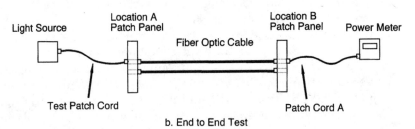

b. End to End Test

Figure 12.4 Power meter test configuration.

record the optical power level as $L_{\text{MeterLoop(dB)}}$ or $P_{\text{MeterLoop(dBm)}}$ for a dBm scale.

6. For the end-to-end configuration using the dB meter reading, the fiber optic total facility loss is the meter's optical loss minus patch cord A's loss:

$$L_{\text{Facility(dB)}} = L_{\text{MeterEnd(dB)}} - L_{\text{CordA(dB)}}$$

If meter readings are only in dBm, the fiber optic total link loss is determined by subtracting the meter reading and patch cord loss from the initial reference value $P_{\text{Ref(dBm)}}$:

$$L_{\text{Facility(dB)}} = P_{\text{Ref(dBm)}} - P_{\text{MeterEnd(dBm)}} - L_{\text{CordA(dB)}}$$

For the loop back configuration using the dB meter reading, add loop back patch cord B's loss to the equation. The result is then divided by two to determine the average loss for one fiber:

$$L_{\text{Facility(dB)}} = \frac{(L_{\text{MeterLoop(dB)}} - L_{\text{CordA(dB)}} - L_{\text{CordB(dB)}})}{2}$$

If the loop back readings are all in dBm, then one fiber's loss is calculated as follows:

$$L_{\text{Facility(dB)}} = \frac{(P_{\text{Ref(dBm)}} - P_{\text{MeterLoop(dBm)}} - L_{\text{CordA(dB)}} - L_{\text{CordB(dB)}})}{2}$$

7. Record your results and the procedure you used.

8. This procedure can be repeated to test all the fibers in the facility.

Example 12.2 A fiber optic facility is being measured for total link loss using the end-to-end test configuration with a power meter in dB (as shown in Fig. 12.5b). The power meter is calibrated to 0 dB using the light source and the test patch cord. Patch cord A was measured prior to the test and was found to have a loss of 0.5 dB. In the end-to-end test configuration (Fig. 12.5b), the power meter loss reading is 8.1 dB. Note that some power meters may show the dB loss value as a negative (-8.1 dB) in the dB scale. The meter derives this value from the gain formula (see Sec. 12.1). It should be converted to the positive loss value (8.1 dB) for use in all calculations. What is the fiber optic facility loss?

Because only one element—patch cord A—is added to this link test, its loss must be subtracted from the meter reading to determine the system loss:

$$L_{\text{Facility(dB)}} = L_{\text{MeterEnd(dB)}} - L_{\text{CordA(dB)}}$$
$$L_{\text{Facility(dB)}} = 8.1 \text{ dB} - 0.5 \text{ dB}$$
$$L_{\text{Facility(dB)}} = 7.6 \text{ dB}$$

Therefore, the fiber optic facility loss is 7.6 dB.

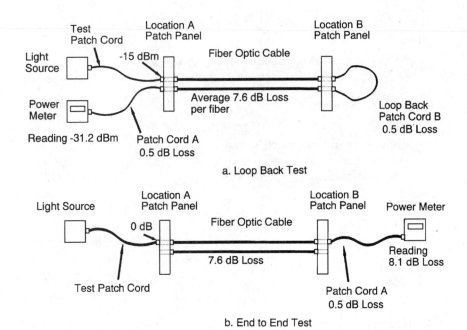

Figure 12.5 Power meter measurement example.

Example 12.3 The system in Fig. 12.5a was measured using the end-to-end configuration with a power meter in dBm. Using the test patch cord, the power meter reference value was recorded to be -15 dBm. Both patch cords A and B were measured prior to the test and were found to each have a loss of 0.5 dB. The power meter reading is -31.2 dBm. What is the fiber optic facility loss?

The loop back dBm formula should be applied:

$$L_{\text{Facility(dB)}} = \frac{(P_{\text{Ref(dBm)}} - P_{\text{MeterLoop(dBm)}} - L_{\text{CordA(dB)}} - L_{\text{CordB(dB)}})}{2}$$

$$L_{\text{Facility(dB)}} = \frac{(-15 \text{ dBm} - -31.2 \text{ dBm} - 0.5 \text{ dB} - 0.5 \text{ dB})}{2}$$

$$L_{\text{Facility(dB)}} = 7.6 \text{ dB}$$

Therefore, the average facility loss is 7.6 dB.

12.5 Return Loss Measurement

A *return loss measurement* is performed if a laser source is used (primarily single mode fiber). The measurement determines the amount of optical power that is reflected back into the laser source by the fiber.

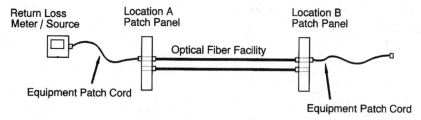

Figure 12.6 Return loss test configuration.

High reflected optical power into a laser source can affect the stability of the laser and can cause erratic operation. Reflected power should be maintained below laser transmission equipment specifications.

To measure the return loss, connect a return loss meter source to the transmit fiber using the installed equipment patch cords (do not use test patch cords). Adjust the meter to the proper settings and record the return loss (see Fig. 12.6).

If the reflected optical power is greater than transmission equipment specifications, try switching to better-quality patch cords and connectors. Also, angle-type fusion splices can help reduce return loss.

12.6   Optical Link Budget

An optical link budget is a tabulation of all the losses (or gains) in a fiber optic link. These losses are due to all components in the optical link, such as cable, connectors, splices, attenuators, and so on. It also includes the light source's average output power, receiver sensitivity, and received optical power.

An optical link budget is normally used during the design stage to determine proper system operating light levels. It can also be used for troubleshooting links or during maintenance procedures to monitor optical degradation.

Optical link budget: An example

(*a*) Optical fiber loss at 1310 nm:
      3.5-km length at 1.8 dB/km   6.3 dB

(*b*) Splice loss:
      2 splices at 0.5 dB/splice   1.0 dB

(*c*) Connection loss:
      2 connections at 1.0 dB/connection   2.0 dB

(*d*) Other component losses   0.0 dB

(*e*) Design margin   2.0 dB

(*f*) Total link loss   11.3 dB

(g) Transmitter average output power	−10.0 dBm
(h) Receiver input power:	
(g minus f)	−21.3 dBm
(i) Receiver dynamic range	−10 to −30 dBm
(j) Receiver sensitivity at BER 10^{-9}	−26.0 dBm
(k) Remaining margin:	
(h minus j)	4.7 dB

Optical fiber link loss (item a in the optical link budget example) is the total optical fiber loss for the installed fiber length at the operating wavelength. This value can be determined from the manufacturer's specifications sheets. It is specified in the attenuation form dB/km at wavelength and should be multiplied by the fiber length in kilometers to obtain the total fiber dB loss. It can also be measured using a power meter and light source or OTDR.

Splice loss (item b) is the total loss due to all mechanical or fusion splices in the fiber optic link. During the design stage, this value is normally estimated using the manufacturer's specifications. Fusion splices are usually below 0.1 dB, and most mechanical splices are below 0.5 dB loss. Splice loss can also be measured using an OTDR.

Connection loss (item c) is the total loss due to all connections in the optical fiber link. Exclude losses due to the connectors that attach the patch cord to the lightwave equipment; these losses have already been factored into the equipment's specifications by the manufacturer. One connection loss results from two connectors being attached together using an adapter. Connection loss can be estimated using the manufacturer's specifications or can be measured with an OTDR.

Other component losses (item d) include the total loss due to other devices in the fiber optic link.

The design margin (item e) is an estimated level of loss that will provide sufficient confidence to achieve proper equipment operation during the life span of the system. Over time, all systems degrade to some degree. This should be anticipated in the design stage so that system life can be maximized. Contributing factors include optical fiber increase in attenuation due to OH ingress, decay of the light source output power, receiver sensitivity decrease, increased link losses due to repair splices, and increased connection losses due to contamination.

The total link loss (item f) is the sum of all optical loss in dB (lines a through e of the optical link budget; see also Fig. 12.7).

Transmitter average output power (item g) is the average optical output power of the lightwave equipment using a standard data test pattern. For an analog system, a standard tone or video test pattern is

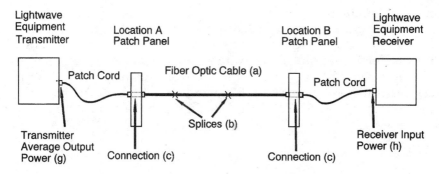

Figure 12.7   Optical link budget losses.

used. This level can be obtained from equipment specifications during the design stage and measured with a power meter after installation.

Receiver input power (item h) is the received optical power at the lightwave equipment. This is calculated by subtracting the total link loss from the transmitter's average output power (line g minus line f of the optical link budget; see also Fig. 12.7).

Receiver dynamic range (item i) is the light-level window in dBm at which a receiver can accept optical power without degrading performance. The optical receiver is designed to accept light within the manufacturer's predetermined levels (measured in dBm). If the light is too strong, the receiver circuit will saturate, and the equipment will not operate. This is normally caused by insufficient fiber link attenuation. To remedy the problem, optical fiber attenuators are inserted in the link until the received optical power is within the receiver's dynamic range. Attenuators can be installed at the equipment connector or at the patch panel connector on the receive-or-transmit fibers. If the attenuators are installed on the transmit fiber, ensure that the return loss measurement is within the equipment's specification. Receiver input power (item h) should be within the receiver dynamic range (item i).

Often lightwave equipment designed for short distance transmissions, the optical receiver is designed to accept the full light source power (no link attenuation) and the minimum light source power (maximum link attenuation). For these cases, fiber attenuators are not required, and this step can be ignored.

Receiver sensitivity at the bit error rate (BER) (item j) is the minimum amount of optical power necessary for the lightwave equipment receiver to achieve the BER in a digital system. In an analog system, it is the minimum amount of optical power necessary for the lightwave equipment receiver to achieve the signal-to-noise (S/N) level.

The remaining margin (item k) is the additional optical loss that the fiber optic link can tolerate without affecting system performance. It is calculated by subtracting the receiver sensitivity at BER from the receiver input power (line h minus line j of the optical link budget; see also Fig. 12.7). This value should always be greater than zero.

13

OTDR Test Procedure

An optical time domain reflectometer (OTDR) is used to obtain a visual representation of an optical fiber's attenuation characteristics along its length. The OTDR plots this characteristic on its screen, in graph format, with the distance represented on the x axis and attenuation represented on the y axis. Information, such as fiber attenuation, splice loss, connector loss, and anomaly location can be determined from this display.

OTDR testing is the only method available for determining the exact location of broken optical fibers in an installed fiber optic cable when the cable jacket is not visibly damaged. It provides the best method for determining loss due to individual splices, connectors, or other single-point anomalies installed in a system. It allows a technician to determine whether a splice is within specification or requires redoing. It also provides the best representation of overall fiber integrity.

In operation, the OTDR sends a short pulse of light down the fiber and measures the time required for the pulse's reflections to return to the OTDR. Fiber imperfections and impurities cause the reflections along the fiber.

Knowing the optical fiber's index of refraction and the time required for the reflections to return, the OTDR computes the distance traveled by the reflected light pulse:

$$\text{Distance} = \frac{3 \times 10^8 \times \text{time}}{2 \times \text{index of refraction}}$$

The OTDR also measures the power of the reflected light pulse and produces an optical fiber attenuation-by-distance display.

The following two sections describe OTDR testing equipment and a general OTDR testing procedure. For additional details, the OTDR manufacturer's manual should be consulted.

13.1 Equipment

OTDR:

- proper wavelengths
- proper connectors
- single mode or multimode fiber sizes
- sufficient dynamic range for length of fiber

Test patch cords and pigtails:

- at least one length as required
- proper connectors
- proper fiber size

Connector cleaning solution, swabs, compressed air

Bare fiber adapter

Index-matching liquid or gel

Cleaver

Cable and fiber strippers

Dead-zone fiber (if required by OTDR)

Before proceeding with OTDR measurements, the OTDR should be checked to ensure that it has sufficient range capability to measure the entire optical fiber length. The following example shows a calculation that estimates the OTDR's fiber range in kms. The combined lengths of the fiber optic cable(s) to be measured should be less than this range.

Example 13.1

   a. Manufacturer's specifications for OTDR's dynamic range is 25 dB [$P_{OTDR(dB)}$].

   b. Optical fiber loss at 1310 nm is 0.48 dB/km, as per manufacturer's specifications.

   c. The average splice loss is estimated at 0.02 dB/km, from three splices in 1-km section.

$$\frac{(0.01 + 0.02 + 0.03)}{3} = .02 \text{ dB/km}$$

   d. Total estimated loss at 1310 nm for a 1 km length is 0.50 dB/km.

$$L_{\text{Total(dB/km)}} = 0.48 + 0.02 = 0.50 \text{ dB/km}$$

e. Approximate OTDR range (dB):

$$d_{R(km)} = \frac{P_{\text{OTDR(dB)}}}{L_{\text{Total(dB/km)}}} = \frac{25 \text{ dB}}{0.50 \text{ dB/km}} = 50 \text{ km}$$

This calculation is performed for all required operating wavelengths.

13.2 Procedure

The following steps should be used to perform an OTDR test:

1. If the optical fiber to be tested is not connectorized, strip the fiber optic cable and expose a 2-m (6-ft) length of the fiber to be tested. Clean and then cleave the fiber to be tested.

2. Connect the OTDR to the fiber to be tested via a pigtail, dead-zone fiber (if required), and bare fiber adapter (see Fig. 13.1). If the optical fiber is connectorized, then connect the OTDR to the fiber by patch cord and dead-zone fiber (if required).

    A *dead-zone fiber* is a 1-km (0.6-mi) length of bare optical fiber with no jacket that is wound on a small reel. It is used for some OTDRs to pass the OTDR's blind spot (dead zone), which can be extended up to 1 km into the fiber under test (dependent on OTDR setup). This would prohibit seeing any type of fiber anomaly in this range.

3. Turn on the OTDR and let it warm up to a stable operating temperature.

4. Enter the proper OTDR parameters for operation, including wavelength, index of refraction of fiber to be tested, and scan mode and resolution.

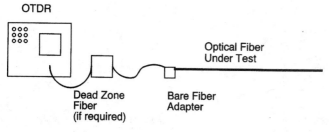

Figure 13.1 OTDR test configuration.

5. Adjust the resolution to display the complete fiber under test. To provide the best resolution, keep the pulse width as short as possible.

6. Measure attenuation for all anomalies, splices, connectors, and overall fiber.

7. Measure the fiber's end-to-end attenuation in dB and dB/km.

8. Repeat steps 1 through 7 for all required optical wavelengths.

9. Record the OTDR's location for these measurements. Print a hard copy of the results or store them on computer diskettes.

10. Repeat steps 1 through 9, with the OTDR connected to the other end of the fiber optic cable. Then average the two results. This will provide a more accurate value:

$$\text{Loss}_{\text{OTDRTotal}} = \frac{\text{Loss}_{\text{DirectionA}} + \text{Loss}_{\text{DirectionB}}}{2}$$

13.3 Determining Anomaly's Physical Location

An OTDR can provide an optical fiber anomaly's location (see Figs. 13.2 and 13.3). However, the anomaly's actual physical location is dependent on the accuracy of the OTDR, the accuracy of the manufacturer's supplied index of refraction (it should be four digits or more) for the fiber core, and the length of excess fiber in a cable.

The excess amount of fiber in a cable is due to the slight bunching of the fiber in the cable tubes and their spiral wrap around the cable's

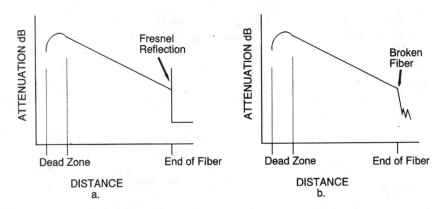

Good fiber trace no splices no anomalies. Clean cleave at end of fiber.

Poor cleave or broken fiber

Figure 13.2 Typical OTDR traces.

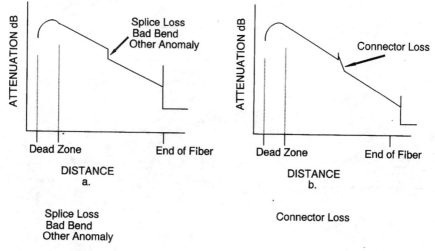

Figure 13.3   Typical OTDR traces.

central strength member. The cable manufacturer specifies this excess amount of fiber in the cable as a percentage of total cable jacket length.

If this percentage of excess is accurate, and the index of refraction is also known to be accurate, the following calculation can be made to determine the physical location of an anomaly:

$$d_{\text{PhysicalBroken}} = \frac{d_{\text{OTDRBroken}}}{(1 + d_{\text{Excess\%}}/100)}$$

$d_{\text{PhysicalBroken}}$ = Actual cable distance to fiber break

$d_{\text{OTDRBroken}}$ = Length of broken fiber measured by OTDR using an accurate index of refraction for the fiber core

$d_{\text{Excess\%}}$ = Excess amount (in percent) of fiber in cable

For a short cable length this may provide sufficient results. Improved accuracy may be obtained using the following comparative method. This method can also be used if an accurate index of refraction or percentage of excess fiber is not known.

Procedure

1. Use an OTDR and measure the distance to a known reference point in the cable, such as a splice or end-of-fiber location and record it as d_{OTDRRef}. An undamaged fiber should be used (see Fig. 13.4).

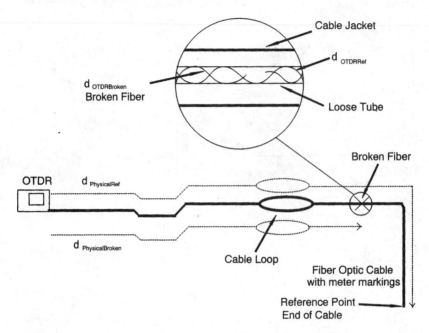

Figure 13.4 Physical fiber break location.

2. Measure the physical cable length to the same reference point as in step 1 using meter markings on the cable only, and record it as $d_{\text{PhysicalRef}}$.

3. Label these distances as follows:

   d_{OTDRRef} = OTDR-measured length of fiber to a reference point

   $d_{\text{PhysicalRef}}$ = Actual physical length of cable determined by cable meter marks to reference point

4. Using the OTDR, measure the broken fiber length (in meters or feet).

   $d_{\text{OTDRBroken}}$ = Length of broken fiber measured by OTDR

5. Calculate the actual physical cable distance to the broken fiber as follows:

$$d_{\text{PhysicalBroken}} = \frac{d_{\text{OTDRBroken}} \times d_{\text{PhysicalRef}}}{d_{\text{OTDRRef}}}$$

This distance is not the route distance but only the cable length distance. Follow the cable jacket length markings to determine the route location.

14

Installation Test Procedure

14.1 Fiber Optic Cable Tests

A fiber optic cable should be tested three separate times during an installation:

1. *The Reel Test.* After the cable is received from the manufacturer and is still on the shipping reel, it is tested for manufacturer's defects or shipping damage. Any anomalies not meeting specifications should be reported immediately to the manufacturer or shipper. The cable should not be installed until it passes this test.

2. *The Splicing-Installation Test.* This test can be performed as soon as the cable is installed in the route, all splices have been completed, and while the splice crews are still on site. It can identify any cable damage resulting from the installation process. It can also provide a splice-loss verification before splice enclosures are permanently mounted.

3. *The Acceptance Test.* This test is performed after the entire fiber optic system is complete and ready for commissioning. It provides the final commissioning data for engineering acceptance and archive data.

14.1.1 Reel test

As soon as the cable reels are delivered each fiber in the reel should be tested with an OTDR (see Fig. 14.1). This will establish that the optical fibers have been received in good condition from the manufacturer and are not damaged. OTDR fiber traces should be recorded and kept on diskette or in paper files.

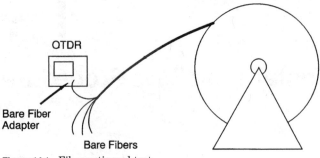

Figure 14.1 Fiber optic reel test.

Procedure

1. Loosen one free end of the fiber cable and strip it to expose all fibers (avoid unlagging the reel).

2. Strip and clean the individual fibers.

3. Using a bare fiber adapter, connect an OTDR to each fiber and record its trace. Use a dead-zone fiber if required. For each fiber, record the following for all required optical wavelengths:

   - total attenuation
   - attenuation per kilometer
   - total fiber OTDR trace
   - any anomalies (zoom in on them and record them)
   - total reel length (marked on reel or obtained from cable markings)
   - total fiber length as indicated by OTDR
   - reel identification number, cable manufacturer, cable type, number of fibers in the cable
   - measurement direction
   - date
   - test equipment and serial numbers
   - crew members

   All anomalies should be reported immediately. Each fiber should be free of anomalies and have no visible splice points.

4. After all fibers are tested, cut off the loose fibers and reseal the cable end to prevent entry of moisture and dirt. Secure the cable to the reel.

Confirm all results with the manufacturer's or engineering specifications. For installations where an OTDR is not available, an optical power meter test should be conducted to confirm fiber attenuation as specified by the manufacturer. Both fiber cable ends are required for this test.

14.1.2 Splicing-installation test

After each cable end is spliced but before the splice enclosures are permanently mounted and while the splice crew is still on site, the installed cable length and splices should be tested (see Fig. 14.2). If only one cable length is installed, this test can proceed immediately after termination of the fiber connection.

Tests are performed on each fiber, at all operating wavelengths, and from both directions.

Procedure

1. Two OTDR crews, in radio contact, prepare the far ends of both cables for testing.

2. The sequence of fibers to be tested is identified and testing is begun. A dead-zone fiber is used if required.

3. Each fiber is tested in one direction, then retested in the opposite direction. The following information is recorded:

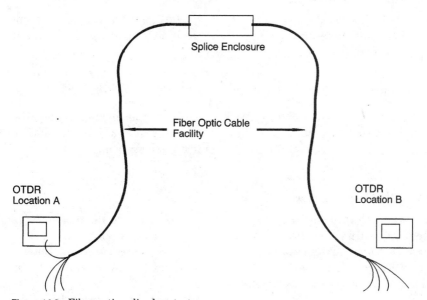

Figure 14.2 Fiber optic splice loss test.

- total length attenuation
- attenuation per kilometer
- total fiber trace
- any anomalies (zoom in on them and investigate)
- splice loss (gain) and splice trace
- total cable length obtained from cable meter markings
- total fiber length as indicated by OTDR
- cable manufacturer, cable type, number of fibers in the cable, cable reel identification number
- measurement direction
- date
- test equipment and serial numbers
- crew members

Splice loss, or gain, and overall cable loss is tabulated for each fiber at each operating wavelength. Splice loss, or gain, from both OTDRs are recorded, then the average is calculated to obtain an overall splice loss. If a gain is recorded, it is entered as a negative.

4. All bad splices should be identified and redone immediately. All fiber anomalies, showing a loss greater than the engineering specification, should be reported immediately.

5. Newly installed cable lengths should have no optical fiber anomalies that have greater loss than specified. Ensure that there are no broken fibers.

6. After all fibers and splices have been tested to engineering specification, securely mount the splice enclosures and cable.

For an inexperienced crew, this test can be performed during the splicing procedure to confirm fusion splicer test results. The optical fibers should be retested once the splice enclosure has been permanently mounted.

14.1.3 Acceptance test

Once installation is complete and the fiber is ready for modem connection, a final acceptance test is conducted between connectors to ensure that the fiber optic link meets the engineering link budget specification. This test is usually conducted by the engineer or by technicians with engineers present.

The following test is done on the complete length of each optical fiber:

1. An OTDR is connected to one end of the fiber optic link.

2. The complete link is scanned, and the trace is stored. The following information is recorded at all operating wavelengths:

   - total length attenuation
   - attenuation per kilometer
   - total fiber trace
   - any anomalies (zoom in on them and investigate)
   - splice loss
   - connector loss
   - total link length obtained from cable meter markings
   - total link length as indicated by OTDR
   - cable manufacturer, cable type, number of fibers in the cable
   - measurement direction
   - date
   - test equipment and serial numbers
   - crew members

3. An optical power meter and source are connected. Power meter readings are taken for each fiber at all operating wavelengths.

4. A return loss meter is connected to record the reflected power of fibers at both equipment terminating ends (if applicable).

5. A power meter measurement of equipment transmitter optical power is helpful for maintenance purposes. The output power of the lightwave equipment is recorded with a common output pattern, such as a digital, all-ones pattern. Lightwave equipment power levels are measured at equipment output and receiver points. Data pattern, location, and levels are recorded.

14.2   Fiber Acceptance Criteria

Optical fiber acceptance criteria should meet engineering specifications.

Maximum optical fiber attenuation in dB at a wavelength:

$$\text{Loss}_{max} = d_{cable} \times L_{dB/km} + (L_{dB/Splice} \times N_{splices})$$

where d_{cable} = Length of cable in kilometers

$$L_{\text{dB/km}} = \text{Manufacturer's fiber loss specification per kilometer at wavelength}$$

$L_{\text{dB/km}}$ = Manufacturer's fiber loss specification per kilometer at wavelength

$L_{\text{dB/Splice}}$ = Maximum loss, per splice, at wavelength

N_{splices} = Number of splices in cable segment being tested

Average attenuation per kilometer at a wavelength:

$$\text{Average}_{\text{dB/km}} = \text{Loss}_{\text{max}}/d_{\text{cable}}$$

Maximum splice loss. As per the engineering specifications, maximum splice loss depends on fiber type, splice type, and wavelength. Generally, for mechanical splice the maximum splice loss is less than 0.5 dB. For fusion splice, it is generally less than 0.1 dB.

Maximum connection loss. Connection loss depends on fiber type, connector type, and wavelength and is generally less than 1.5 dB. A good connection average loss would be about 0.5 dB or less.

14.3 Bit Error Rate Test (BERT)

A *bit error rate test (BERT)* is a point-to-point test performed to determine the quality of a digital system or channel (see Fig. 14.3). This test is generally conducted after all equipment has been completely installed and is in full operational mode but before actually placing the system into service. It can also be used to test an existing communication system or channel; however, the system or channel must be taken out of service to perform the test.

The test measures how many data bits in a transmission are received in error. The measure is a bit error rate (BER), which is the ratio of bits received in error to the total number of bits transmitted:

$$\text{BER} = \frac{\text{Number of received bits in error}}{\text{Total number of bits transmitted}}$$

A BER of 10^{-9} indicates that on average one error bit is received for every billion sent. A BER of 10^{-9} is adequate for many installations, including voice communications. For high-quality fiber optic systems, a BER of 10^{-11} is common.

The following procedure can be used to perform a BERT on a communication channel. The channel must be placed out of service for this test to be performed.

Procedure

1. The communications channel or system to be tested is placed out of service. DTEs at both ends of the channel are disconnected.

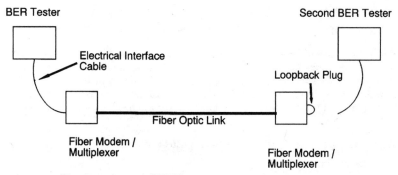

Figure 14.3　Bit error rate test (BERT).

2. Two BER testers are connected to both ends of the channel using standard electrical interfaces such as RS232, RS449, V.35, T1, and so on.

3. If only one BER tester is available, the opposite end of the link is physically looped back with a loop back plug to complete the signal test path. If a loop back plug is not available, then a software loop back is used (however, the entire channel to the connector interface is not tested).

4. Both testers are configured and set to transmit a pseudorandom test pattern that will simulate live traffic, such as "2047 or $2^{23}-1$."

5. The test is started and run for a specified period of time to determine the BER of the link. The testing period should be long enough to provide reasonable confidence of link performance. Link tests over one or two days are common for high data rate links. An event timer can be used to log erratic errors.

6. After the test period, the recorded BER should meet or exceed the system's BER specification.

Bit error rate test: An example　A newly installed fiber optic T1 DS1 data communications link is to be tested before being placed into service. If the manufacturer specifies its communication equipment will provide a BER rate equal to or better than 10^{-10}, can this be tested?

　The claim can be verified during a test period. After the equipment has been completely installed and is operational, two BER testers are connected to the channel to test the manufacturer's claim. The BER testers use the standard four-wire DS1 interface to connect to the communication channel. The BER testers are properly configured and set to test the channel using the "QRSS" pattern. The T1 data rate is 1.544 Mbps.

　The BER claim is 10^{-10} or, on average, 1 bit error for every 10^{10} bits transmitted. The amount of time required to transmit 10^{10} bits is as follows:

Required time = Number of bits to be transmitted/Data rate

Required time $= (10^{10}/1.544) \times 10^6$

Required time $= 6476$ seconds or 1.8 hours

For a data rate of 1.544 Mbps, the transmission must last for at least 6476 seconds or 1.8 hours for 10^{10} bits to be transmitted. The duration of the transmission can be doubled or tripled to increase the confidence level.

It was decided to run the test for 5.4 hours (19,428 seconds). After the test was completed, a total of two errored bits was recorded. The tested BER is then as follows:

Total bits transmitted $= 19,428$ seconds $\times 1.544$ Mbps

Total bits transmitted $= 2.9997 \times 10^{10}$

BER = Number of received bits in error/Total number of bits transmitted

BER $= (2/2.9997) \times 10^{10}$

BER $= 6.667 \times 10^{-11}$

During the test interval, the BER was therefore better than the manufacturer's specifications.

14.4   Receiver Threshold Test

A *receiver optical power threshold test* is conducted to determine the lightwave receiver optical power threshold (see Fig. 14.4). This is the minimum light signal power required at the optical receiver to meet the equipment manufacturer's BER specifications for a digital system or the signal-to-noise (S/N) specifications for an analog system. This value can be used to monitor a receiver's performance and to determine the link's *optical margin.*

The following procedure can be followed to measure the receiver optical power threshold:

Procedure

1. Note the equipment's BER for a digital system or the S/N value for an analog system—as well as the receiver's threshold values— from the manufacturer's specifications.

2. Take the lightwave equipment out of service for this test. For a digital system, connect a BER tester to the lightwave equipment and configure it for a standard BERT. For an analog system, set up an S/N test.

3. Conduct the BERT using a pseudorandom pattern, and record the system's BER. For an analog system, conduct the S/N test and measure the S/N.

4. Disconnect the patch cord at the lightwave receiver and connect it

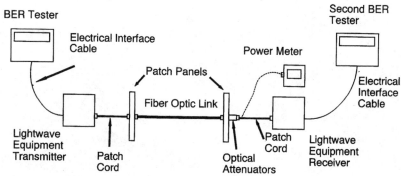

Figure 14.4   BER optical threshold test.

to an optical power meter. Measure the received operational average optical power in dBm for the measured system BER or S/N value.

5. Reconnect the patch cord to the lightwave receiver.

6. Add optical attenuators to the optical receiver link until the BER or S/N drops to the manufacturer's minimum specifications.

7. Disconnect the patch cord at the lightwave receiver and connect it to an optical power meter again. Measure the received test average optical power in dBm for the minimum BER or S/N value. This is the receiver optical power threshold, which should be compared to the manufacturer's specifications. Any further reduction in optical signal will degrade the system's performance below the manufacturer's specifications.

8. For a full duplex link, conduct this test in one direction and then repeat it again in the other direction to determine threshold values for both lightwave optical receivers.

The optical margin in dB is the total attenuation added to the operational link to lower the BER or S/N value to the manufacturer's minimum specifications.

15

Lightwave Equipment

Lightwave equipment is a general term used here to refer to any optical fiber terminating equipment that converts electrical signals into fiber optic light signals and fiber optic light signals back to their original electrical form. This equipment includes fiber optic converters, repeaters, or modems. The equipment is usually simple in design; easy to install, with little required configuration; and easy to maintain. Lightwave equipment can also be combined with other communication equipment to provide a more integrated unit. This type of equipment is commonly referred to by its main function and has plug-in modules that provide the optical-to-electrical conversion. Two popular equipment types are multiplexers and LAN hubs. Both are multifunction units with available optical-to-electrical conversion modules.

15.1 Optical Modem

For installations that require that a few signals be transmitted over a distance, an individual optical modem set can be implemented (see Fig. 15.1). Optical modems are available for most communication signals. One modem set and optical fiber pair (two fibers—one transmit and one receive) is used for each signal that needs to be transmitted. The advantage of this configuration is that the optical modems are relatively inexpensive, easy to use (many require little or no configuration), and available for many applications. The disadvantage of the configuration is the need for dedicated fibers or fiber pairs for each signal. Optical fibers can be used up quickly in a cable. Most optical fiber links use two fibers to provide full two-way communications. For

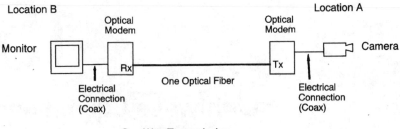

a. One-Way Transmission

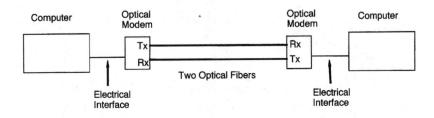

b. Two-Way Transmission

Figure 15.1 Optical modem link.

some systems, such as video links or public address, voice-only, one-way communication is required, and therefore one fiber is used. The following table illustrates the various applications of links with one, two, and four fibers.

Application	One Fiber	Two Fibers	Four Fibers
Voice (one-way, PA)	*		
Voice (telephone)		*	
Video (security)	*		
Video (interactive)		*	
Control Systems (PLC)		*	
Telemetry	*	*	
Data communications		*	
Multiplexer		*	
Ethernet		*	
Token ring		*	
FDDI		*	*
Sonet		*	*

New technologies are available that can combine both transmitting and receiving optical signals onto one fiber. However, they are not widely used because of their added loss to the fiber link and their cost.

15.2   Multiplexer

A *multiplexer* installation allows a number of signals to be combined onto a fiber pair. This reduces the number of fibers required in an installation, which can offer a significant cost advantage for long distance installations. Most multiplexers also have some sort of network management capability. This allows an operator to monitor multiplexer performance from a central location. Multiplexers tend to be expensive and complicated to install. Personnel operating this equipment may require special proprietary training from the manufacturer. Such equipment is commonly used by large communication carriers such as telephone companies.

As shown in Fig. 15.2, multiplexers have the capacity to combine a number of signals into a two-fiber transmission link. The more elaborate types can combine different signal types such as voice, video, and data into one aggregate. Other multiplexers can only offer either a data or voice interface. Each communication signal is connected to a multiplexer electrical interface specific to the signal type. Video signals use coaxial cable. Point-to-point computer signals can use a number of interfaces, including RS232, RS449, or V.35. Telephone connections can be made with standard two-wire POTS interface, E&M signaling, or a T1-DS1 connection.

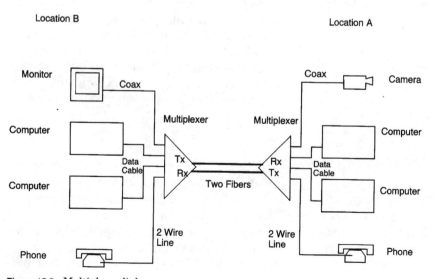

Figure 15.2   Multiplexer link.

15.3 Optical Amplifiers

Optical amplifiers are used to achieve long distance transmissions of over 200 kilometers before signal regeneration is required. They are used on single mode fibers, with laser sources, and operate in the 1310- or 1550-nm wavelengths. Optical amplifiers are special types of lasers that directly boost optical signal strength without converting the signal into an electrical form.

> *Optical output from a power amplifier is very strong, and extreme care should be exercised when working with this equipment to prevent eye damage. All equipment should be turned off before disconnecting optical fibers. Never turn equipment on when all fibers are not properly connected to equipment.*

Electrical power to the amplifier should only be turned on after all fiber connections have been properly completed. Optical amplifiers should have a safety interlock that turns off the laser optical output if the fiber link is broken or disconnected.

When testing a fiber optic link, with OTDR or power meter, that has an optical amplifier installed, the optical amplifier should be turned off and disconnected from the fiber optic link. Each segment of the fiber optic link should be tested separately. Since an optical amplifier will only permit optical signals to pass in one direction, an OTDR requires bidirectional signal flow in a fiber (primary and reflected light pulse) and, therefore, it will not work.

Optical amplifiers are available for connection at the lightwave equipment locations to provide a powerful boost to the optical transmission. They are also available for installation at mid-span locations for the amplification and boosting of weak optical signals along a long fiber optic route (see Fig. 15.3). Some mid-span amplifiers do not regenerate the signal (they only amplify it), therefore pulse distortion due to fiber dispersion should be carefully managed. Mid-span location amplifiers normally require a small, environmentally controlled hut or enclosure for the amplifier.

15.4 Light Sources

There are two types of light sources used by lightwave equipment for optical fiber transmission, *light-emitting diodes* (*LEDs*) and *lasers*.

LEDs are economical and are common for short distance, low data rate applications. They are available for all three wavelengths but are most common at 850 and 1310 nm (850 nm LEDs are usually the least expensive). Light power from an LED covers a broad spectrum,

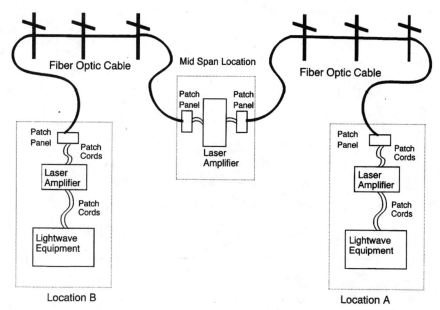

Figure 15.3   Laser amplifier system.

from 20 to over 80 nm (see Fig. 15.4). The LED is more stable and reliable than a laser in most environments.

Lasers are more expensive. The advantages of using a laser are in the high modulation bandwidth (over 2 GHz), with high optical output power and narrow spectral width. Their application is in long distance, high data rate requirements. Lasers are common in single mode optical fiber applications, and their light power covers a very narrow spectrum, usually less than 3 nm. This results in a low chromatic dispersion value and hence high fiber bandwidth. Their life span is shorter than that of an LED. Lasers are sensitive to the environment (especially to temperature variation) and to such fiber design parameters as reflected optical power.

> *Optical output from a laser is strong and can easily damage the eye. Never look into laser light or a fiber coupled to a laser. Ensure that all laser sources are powered off before disconnecting the fibers. Care should be exercised when working with laser sources.*

15.5   Optical Detection

Optical detection occurs at the lightwave receiver's circuitry. The photodetector is the device that receives the optical fiber signal and con-

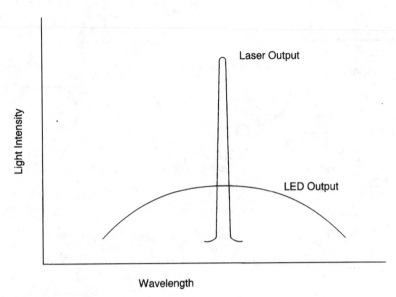

Figure 15.4 Light source spectrum.

verts it back into an electrical signal. The most common types of photodetectors are the *positive intrinsic negative photodiode* (*PIN*) and the *avalanche photodiode* (*APD*).

PIN photodiodes are inexpensive, but they require a higher optical signal power to generate an electric signal. They are more common in short distance communication applications.

The APD photodiodes are more sensitive to lower optical signal levels and can be used in longer distance transmissions. They are more expensive than the PIN photodiodes and are sensitive to temperature variations.

Both photodiodes can operate at similar, high-signal data rates.

Some receiver photodetector circuits operate within a narrow *optical dynamic range*.

> The receiver's optical dynamic range is the light-level window in dBm through which a receiver can accept optical power.

The receiver will only accept light within the manufacturer's predetermined levels (measured in dBm). If the light is too strong, the receiver's circuit will saturate, and the equipment will not operate. This is normally caused by insufficient fiber link attenuation. To remedy this problem, optical fiber attenuators are inserted in the link until the received optical power is within the receiver's dynamic range. Attenuators can be installed at the equipment connector or at the patch panel connector on the receive or transmit fiber. If installing on the transmit fiber, ensure that return loss measurement is within the equipment's specifications.

If the light is too weak, the receiver will not be able to detect the signal, and the equipment will not operate. In this case, attenuation must be removed from the fiber optic link. Many receivers designed to operate for short distance communication, such as for LAN communication, are designed not to saturate when full light power level is received (no link attenuation).

16

System Integration

16.1 Office Installation

Figure 16.1 shows a typical office building installation. Lightwave equipment in building A is located in a secure telecommunication closet. The equipment is mounted in standard 19-in cabinets (inside mounting width) or in simple equipment racks. Sizes of 23 and 24 in

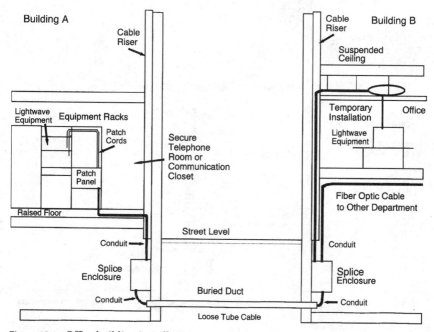

Figure 16.1 Office building installation.

are also common for larger equipment. Cabinets provide for a totally enclosed, neater-looking installation. They should have at least a front and back door to provide adequate equipment and cabling access. A smoked front glass door will allow the equipment's light status to be monitored without opening the door. In a busy room, these doors can be equipped with key locks that can provide additional security. In Figure 16.1, the center rack is assigned to lightwave equipment installation; the right rack is assigned to fiber optic cable patch panels. The example in Fig. 16.1 has lightwave equipment and patch panels assigned to separate racks, which is common for a large installation that requires plenty of room for expansion. For lesser systems, both lightwave equipment and patch panels can be mounted in the same rack. Figure 16.2 shows a typical rack/cabinet layout with lightwave equipment and the patch panel mounted together.

In building A in Fig. 16.1, two fiber optic patch cords connect the lightwave equipment to the patch panel mounted in the adjacent rack. Most lightwave equipment uses two optical fibers for *full duplex* communication, one fiber for the transmit signal and the other fiber for the receive signal. Lightwave equipment that is capable of full-duplex transmission on one fiber is available, but it is not popular.

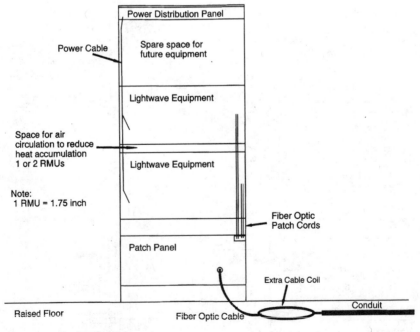

Figure 16.2 Rack/cabinet layout.

Computer equipment located in work areas connect to the telecommunication closet and lightwave equipment by horizontal cabling. Horizontal cabling can be fiber optic or copper.

A fire-rated, tight-buffered indoor fiber optic cable runs in a conduit between the patch panel and the basement splice enclosure. An oversized conduit can be installed to accommodate future cables. All conduit bends must be smooth, with a radius larger than the fiber optic cable's minimum bending radius for the loaded condition.

The tight-buffered cable is spliced to the outdoor loose tube cable in the basement splice enclosure. The splice enclosure has the capacity to accommodate a number of cables to allow the fiber to be branched to different locations in the building. A patch panel can also be used in this location instead of the splice enclosure. The patch panel will provide a fiber intermediate cross-connect facility, and allow for easy future changes.

The outdoor loose tube fiber optic cable is routed through buried ducts into building B. The loose tube cable is spliced to two tight-buffered cables servicing two different departments. A third outdoor loose tube cable could also have easily been spliced here to continue the cable link to a third building. Again, a patch panel could have been used instead to provide an intermediate cross-connect facility.

The tight-buffered cable continues to an upper floor through conduits in the building's cable riser. It is then routed along a suspended ceiling to a temporary equipment installation. An additional length of coiled cable is placed in the ceiling to allow the equipment to be moved to a permanent telecommunication closet some time in the future. The amount of cable in the coil depends on the distance of the route to the permanent installation as well as the amount of cable needed for splicing in a patch panel. The entire cable length to the cable coil is protected by conduits.

The cable drops from the ceiling to the lightwave equipment along a supporting pole. Because of the temporary nature of the installation, a patch panel is not used. Connectors are installed directly onto the cable's buffered optical fiber.

16.2 Industrial Plant Installation

Heavy industrial environments require special consideration, and all components should be designed to meet the harsher conditions. Fiber optic cable routing throughout a plant should always be protected by conduits or heavy armored jacket. Cable temperature specifications should exceed all anticipated environmental variations. Outdoor cable installations may require additional protection from the environment or from the movement of heavy equipment. Single-armor or double-armor cables with heavy polyethylene jackets are now available.

As Fig. 16.3 shows, fiber optic cable can be used to provide communications to a remote plant. Aerial cable installations are common because of the available electrical pole line route. The fiber optic cable is placed as high as possible (in conformance with all electrical codes) on the pole to provide maximum ground clearance. At locations where high-load road traffic is anticipated, the cable can be buried in a duct. A 10-ft heavy steel guard can be placed around the cable at the pole riser to protect the cable from accidental damage by vehicles. The cable is routed in conduits and cable trays to the electrical rooms at both locations. Here the cable is terminated in a dust-free enclosure or in other enclosures properly rated for such installations.

Figure 16.4 shows an example of an industrial fiber optic cabinet layout. The cabinet is used to house fiber cable terminations, light-wave equipment, and to store excess fiber optic cable. A typical size is 1.8 m tall by 0.6 or 0.9 m wide and 0.45 m deep (6 ft tall by 2 or 3 ft wide and 18 in deep). Construction is to industry code, commonly consisting of heavy galvanized steel with a solid-hinged sealing door. A door lock is often used to discourage unauthorized entry.

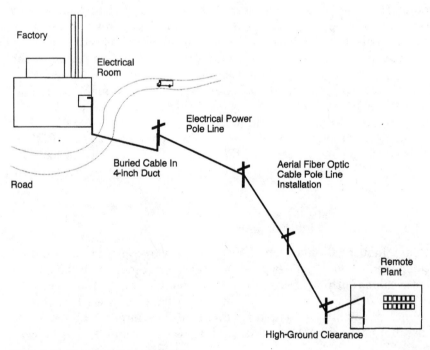

Figure 16.3 Remote plant installation.

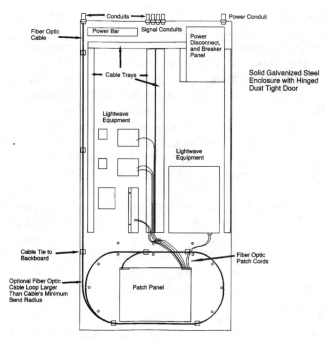

Figure 16.4 Industrial fiber optic cabinet layout.

The cabinet's equipment layout allows room for patch panels, terminating equipment, and power supplies, electrical power disconnects and breaker panels, the test equipment's electrical power bar, and various wiring troughs. Storage of extra fiber optic cable can be accommodated in a cable loop at the bottom of the cabinet. Cable loop minimum bending radius (for static installation) should always be observed. Cable conduits enter the cabinet from the top or bottom and are properly sealed.

Fiber optic patch cords can be routed in dedicated cable trays to lightwave equipment. Lightwave equipment is the wall-mounted type and is secured to the cabinet's back panel. Additional space for associated lightwave equipment and power supplies may be required. Power and signal wiring is placed in separate cable trays and routed to appropriate equipment in other cabinets. Center cable trays can be eliminated in order to mount large equipment.

16.3 Optical Modem System

An optical modem link can be implemented for a point-to-point network. Each signal that needs to be transmitted is connected to a dedicated optical modem set. Modems are available as "standalone" types, external to the equipment, or as the internal card type that can fit into existing equipment slots (i.e., computer slots).

Figure 16.5 provides an example of a modem link. Two computers, a telephone, and a video link use separate optical fiber modems connected through patch cords to a single fiber optic cable. Each modem is independent, that is, the failure of one modem set will not affect other modems. Modem installation is usually quick and straightforward. Most just need to be powered on to be in full operational mode. Communication equipment troubleshooting is also simplified. One or both optical modems can be quickly replaced if suspected of failure. The system can be easily expanded by adding new modems to the fiber optic cable, assuming additional fibers are available. As Fig. 16.5 shows, this configuration can be implemented for a simple point-to-point network. However, if many signals are to be transmitted over a longer distance, a more sophisticated approach should be considered (possibly including a multiplexer) that will reduce the optical fiber demand and the cost of fiber optic cable.

16.4 Multiplexer System

Multiplexer systems make possible the efficient use of both available optical fibers and bandwidth over fibers. A large number of similar or

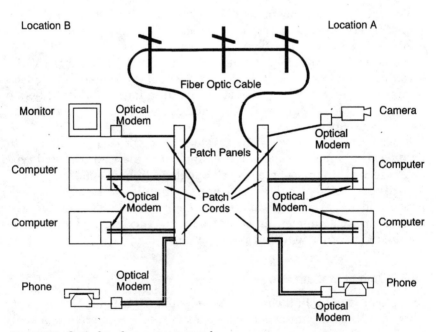

Figure 16.5 Optical modem system example.

different signals can be placed onto two optical fibers. One such system using the Sonet protocol can multiplex 32,256 telephone conversations onto two fibers. As Fig. 16.6 shows, the multiplexer is connected to the fiber optic cable via two patch cords and a patch panel connection at both ends of the link. Equipment such as computers, telephones, and other signals can connect to the multiplexer using proper electrical interfaces.

This system, although more complex to install and maintain, has substantial advantages for some applications. It allows a large number of devices to communicate a long distance using only two optical fibers, thereby saving significant amounts on cable costs. Multiplexers can also provide powerful network management functions. Nodes across the country can be monitored and controlled from a central network control center. Multiplexer circuits can be added, disconnected, tested, monitored, or cross connected from the network control center without dispatching personnel to the multiplexer locations.

16.5 Ethernet

Fiber optics can be used to extend Ethernet LAN distances farther than with conventional cabling, such as 10Base5, 10Base2, or 10BaseT. Coaxial, 10Base5 cabling is limited to 500 m (1640 ft). With the addition of a fiber optic repeater, this distance can be extended to

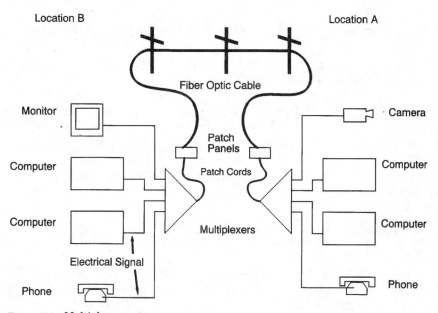

Figure 16.6 Multiplexer system.

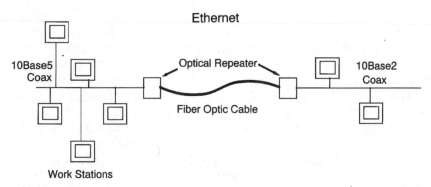

Figure 16.7 Ethernet system.

2 km (depending on network delays). Fiber-type 62.5/125 μm and ST connectors are common for these installations. The use of fiber optic bridges can also extend the distance to over 30 km on single mode fiber (see Fig. 16.7).

16.6 FDDI

Fiber distributed data interface (FDDI) is a standard for a high data rate (100 Mbps) area networking technology that uses fiber optics as the transmission medium. The system is configured in a dual ring topology (similar to a token ring), using a pair of optical fibers, to connect each device on the ring. It is often used as a high-speed campus backbone, as a gateway to lower data rate LANs, or for connection between mainframe computers. As a data highway, it has plenty of bandwidth to carry large amounts of data and is robust enough to provide good fault tolerance and redundancy (see Fig. 16.8).

This dual ring architecture provides a high degree of reliability and security. Under normal operating conditions, the primary ring is used to carry data, while the secondary ring remains idle. In the event of a fault, a break in the fiber optic cable, or if a device fails, stations on both sides of the fault detect and automatically loop back the primary fiber onto the secondary optical fiber. This bypasses the fault condition and recovers the ring operation (see Fig. 16.9).

Each station on the ring can be up to 2 km apart when using FDDI-grade optical fiber. The system can support up to 500 stations on the ring. Maximum ring length is 100 km (according to FDDI Standards). New proprietary interfaces using single mode fiber can increase these lengths; however, individual product manufacturers should be consulted. FDDI standard fiber used for installation is multimode 62.5/125 μm FDDI grade operating at 1310 nm. Standard optical

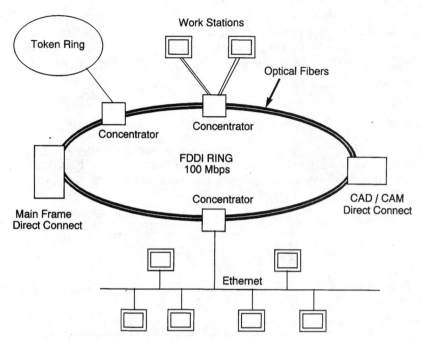

Figure 16.8   FDDI backbone network.

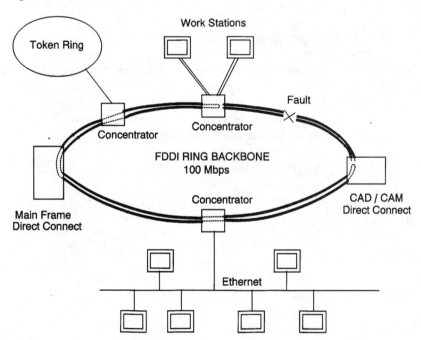

Figure 16.9   Fault in ring.

fiber connectors are the dual-fiber, keyed FDDI type. Some equipment can also accept other connectors such as the ST, FC, or SC type.

Communication is by timed token-passing protocol. This eliminates any delays that could be caused by packet collisions, as occurs in Ethernet.

16.7 Sonet

Synchronous optical network (*Sonet*) is a protocol standard used for fiber optic communication, interfaces, and management. It is defined in eight data rate levels as shown in the following table:

Level	Data rate (Mbps)	Number of voice channels*
OC-1	51.84	672
OC-3	155.52	2016
OC-9	466.56	
OC-12	622.08	8064
OC-18	933.12	
OC-24	1244.16	16,128
OC-36	1866.24	
OC-48	2488.32	32,256

*Using standard, not compressed, 64 kbps channels. This amount can be increased using voice compression techniques.

Sonet provides an international-standard family of transmission channels at optical rates. This results in operability at an optical level between lightwave equipment vendors. It also provides synchronous multiplexing and allows efficient drop-and-insert channel operation at multiple sites.

Because of the high data rates involved, Sonet commonly uses single mode fiber for its communication medium.

17

General Installation Procedure

The following general procedure can be followed for many fiber optic installations.

1. Complete the fiber optic system engineering design and identify the design features, including the following:

   - Identify the exact fiber optic cable route and ensure that it meets all installation specifications. Obtain fiber optic cable installation authorization where required along the route.

   - Determine fiber type—multimode, step index, graded index, or single mode—and diameter.

   - Determine cable type—loose tube, tight buffered, and number of fibers.

   - Ensure that the cables and equipment are designed for the environment. Select outdoor or indoor cable with proper fire rating and jacket material.

   - Determine the optical fiber's connector type and installation procedure (pigtail or direct fiber connection).

   - Determine the field locations and installation requirements for the patch cords, pigtails, patch panels, connectors, and splice enclosures.

   See Chap. 5 for additional details.

2. Identify all safety concerns.

3. Ensure that proper installation and test equipment are available. Make sure that all personnel are properly trained to handle fiber

optic cable and equipment. Determine the correct installation procedure.

4. Order fiber optic cable(s) and all equipment required, as per design, and allow adequate delivery time.

5. Once cable(s) are received, perform the fiber optic cable reel test prior to cable installation (see Sec. 14.1).

6. Prepare the fiber optic cable route and install all conduits, ducts, innerducts, messenger cables, and so on, as required.

7. Install fiber optic cable as required according to engineering design (see Chaps. 3 to 9).

8. Splice all separate cable lengths together, as required, and test complete fiber optic cable link (see Sec. 14.1).

9. Terminate fiber optic cable in appropriate patch panels or splice enclosures and complete cable installation and testing.

10. Test complete fiber optic facility (see Sec. 14.1).

11. Install all terminating lightwave equipment, modems, multiplexers, and so on.

12. Connect terminating equipment to fiber optic facility and test (BERT or equivalent).

13. Record all required installation details (see Chap. 19).

14. Prepare maintenance and repair plans (see Chaps. 18 and 19).

18

Maintenance

Regular maintenance should be performed on a fiber optic system to ensure proper reliable operation. Most maintenance work can be conducted without effecting system operation. Other work, such as optical power level measurement, requires service interruption.

18.1 Nonservice-Effecting Maintenance

The fiber optic system should be visually inspected, at least annually. Patch cord and cable bends should be checked to ensure that the minimum bending radius is not compromised.

Patch cords should be neatly stored or secured in cable trays. Do not tie wrap or bend too tightly. They should never be left dangling.

Fiber optic cable bends should be checked to ensure that they are not too tight. Cable jackets can be inspected for damage, such as cuts, tears, or deformations.

Lightwave equipment and patch panel location should be checked for cleanliness. Cabinet doors should remain closed at all times. Dust should not be permitted to enter or accumulate on equipment. In dusty environments, special sealed cabinets should be used.

Optical attenuation in nonoperational optical fibers (spare fibers) can be measured using a light source and power meter and can be compared to installation records to determine optical fiber deterioration. Under normal operating conditions, fiber attenuation should remain constant over many years.

Lightwave equipment's laser biasing level can usually be checked without affecting equipment operation (consult the manufacturer's specifications). Manufacturer's specifications can also be consulted to determine the proper operating levels.

Pole line installations should be visually inspected for any damage to the cable, messenger, or supporting structure. Messenger and fiber optic cable sag should also be checked. Overhead cables are susceptible to environmental damage, such as wind or ice, as well as damage from birds, rodents, humans, or gunfire.

Underground cable vaults should be checked for cable and support integrity and for rusting. Cable damage could occur from installation of other parties' cables; from damage by rodents, water, ice, and fire; and from cable support corrosion.

All duct, innerduct, and conduit seals should be checked for integrity. Water should not be allowed to enter any fiber optic duct, innerduct, or conduit assembly.

18.2   Service-Effecting Maintenance

Service-affecting maintenance can be performed as frequently as is indicated by the manufacturer's specifications. Usually, equipment is stable for long periods.

Operational fibers can be disconnected and fiber attenuation measured using OTDR and power source/meter. Results are then compared to records to determine any increase in attenuation. Lightwave equipment's optical output power can be measured and compared to records to determine laser or LED aging.

Optical fiber's reflected power can be measured to ensure stable laser operation (single mode fiber only). Receiver optical threshold and BER can also be checked and compared to records.

19

Repair

Repair of broken cables should be carefully planned.

Cable should not be repaired in a wet environment. Cable under repair should be grounded at all times.

The exact location of the fiber break should be determined by visual inspection and OTDR measurements. Usually, sufficient slack is not available in an installed cable to perform a splice without the addition of a patch cable length. A patch cable will introduce two splice losses, as well as patch cable attenuation, into the link budget. The system link budget should be reviewed to ensure that splicing with a patch cable is possible. If it is not, replacement of the entire cable may be required (see Fig. 19.1).

Once the damaged section of cable is identified, it should be cut out. If immediate fiber optic cable restoration is required, a short length of a rugged cable can be laid on the ground and temporarily spliced (mechanical splices can be considered if the link budget is adequate and temperatures do not fall below freezing). A proper rugged style cable can withstand abuse, such as vehicle crushing, tangling, freezing in ice, and so on.

Procedure

1. Determine the exact location of the fiber break and the proper length of patch cable required for splicing at ground level. For outdoor repair, splicing is preferably done in a covered enclosure, such as a van or tent.

2. Determine the mounting locations of splice enclosures. For outdoor repairs, splice enclosures should be outdoor and weather resistant,

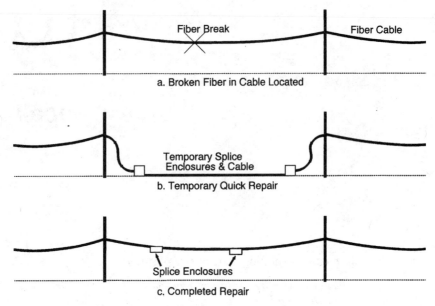

Figure 19.1 Aerial fiber optic cable repair.

with a watertight seal. For pole line repairs, splice enclosures are available for mounting on a messenger or on the pole.

3. Ensure that all repair hardware is available on site, including test equipment, cleaning equipment, cable stripping equipment, mechanical or fusion splicers, and so on.

4. Ensure that all communication in the cable is terminated.

5. Turn off and lock out all fiber optic cable light sources and label them "Out of Service, DO NOT OPERATE."

6. Once the splice enclosures and splicing location have been prepared, cut out the damaged section of cable. Use standard splicing procedures, and splice in the new patch cable.

7. Test all fibers and all splices.

8. After cable splicing and testing are complete, the splice enclosures are mounted to their permanent locations. For a pole line repair the splice enclosures can be mounted on the messenger or on the pole (see Figs. 19.2 and 19.3). Cable slack should be stored in a folded-over manner on the messenger or in a coil on the pole, making sure to observe the cable's minimum bending radius. Make sure that cable twisting does not occur.

9. Splices, and all fibers, are acceptance tested with an OTDR and power meter.

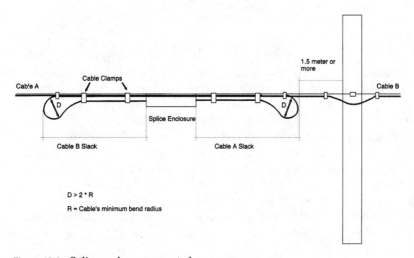

Figure 19.2 Splice enclosure mounted on messenger.

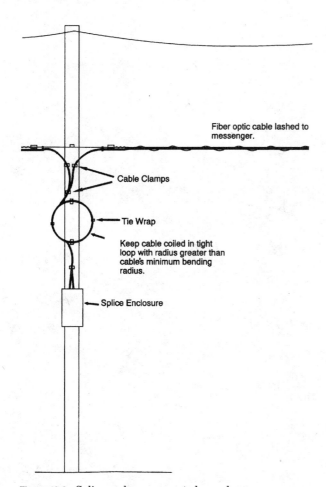

Figure 19.3 Splice enclosure mounted on pole.

Figure 19.2 shows a typical pole line splice enclosure installation on a messenger cable. Excess fiber optic cable length is stored folded over on the messenger. Minimum fiber optic cable bending radius should not be compromised. Splice enclosure is weather proof, watertight, and has special mounting brackets for mounting the messenger.

If mounting on the messenger is not desired, splice enclosures can also be pole mounted. Figure 19.3 shows a typical pole mount configuration. The fiber optic cable coil is placed directly above the splice enclosure. Make sure the cable's minimum bending radius for unloaded conditions is not compromised. Pole-mounting containers are available that will house both the cable coil and the splice enclosure. Enclosures can be mounted high to discourage unauthorized entry.

20

Records

Accurate installation and maintenance records should be kept for future repair or system modifications.

Installation records

1. Installed optical fiber OTDR test documentation including

   - total length attenuation
   - individual fiber traces for complete fiber length
   - paper or computer disk records of all traces
   - losses of individual splices and connectors
   - losses of other anomalies
   - wavelengths tested and measurement directions
   - the manufacturer, model, and serial number of the test equipment

2. Installed power meter test documentation including

   - total link loss for each fiber
   - source optical output power using standard transmission pattern, such as all ones
   - receiver optical threshold level
   - optical margin level
   - return loss measurement (laser sources only)
   - wavelengths tested and measurement directions

- manufacturer, model, and serial number of test equipment

3. As-installed drawings including

- fiber optic cable route drawings and details
- splicing locations
- optical fiber connection diagram
- optical fiber assignments at patch panels and terminations
- connector type
- optical fiber assignments at splice locations
- installed fiber optic cable length
- date of installation
- installation crew members

4. Fiber optic cable details including

- manufacturer of fiber optic cable
- cable type, diameter, weight
- jacket type, fire rating, indoor, outdoor
- manufacturer of optical fiber
- optical fiber type
- optical fiber core and cladding diameter
- optical fiber attenuation per kilometer, and
- optical fiber bandwidth and dispersion details

5. Lightwave equipment specifications including

- manufacturer, model, serial number
- detailed operation, maintenance, and repair instructions
- lightwave specifications, output power, receiver sensitivity, source type, operating wavelength, and so on
- other equipment details

Maintenance records

1. Date, crew members, and test equipment used to test and inspect fiber optic system.
2. Complete power meter and OTDR testing as shown in installation records. Readings should be compared to installation records. Tests are service affecting.

3. Cable route inspected and notes on condition of cable and support structure.

4. Patch panel and splicing assignments checked.

5. Detailed notes on any problems encountered.

21

Troubleshooting

A methodical step-by-step approach should be used to isolate a problem. If the system is operational, all users should be informed of possible disturbances or outages. A log should be kept noting symptoms of the problem and the steps used to identify and rectify it.

Nonservice interrupting

1. Visually inspect all patch cords and patch panels to ensure that the cabling is correct and connected as required.

2. Check to ensure that all equipment has power and is in full, proper operational mode. Check all visual hardware and software indicators for proper configuration.

3. Ensure that patch cords and cables have not been bent tighter than their minimum bending radius.

4. Visually inspect the entire cable route for damage, tight bends, or abnormal conditions.

5. Ensure that all connections are secure. If the problem is intermittent, gently wiggle the cables and connectors to see if the problem occurs (this is possibly service interrupting). Some connectors such as the Biconic type are susceptible to vibrations.

Service interrupting. Before disconnecting optical fibers, ensure that all fiber optic light sources are turned off.

1. Disconnect all patch cords and clean all connectors thoroughly.

2. Test all patch cords. Replace if any are doubtful.

3. Measure attenuation of all fibers with a stable light source and power meter and compare to records.

4. Test all fibers with OTDR and compare results to records.

5. Connect power meter directly to the lightwave equipment light source and measure optical output power. Use standard data transmission test pattern (as shown in installation records), such as all ones. Compare to records.

6. Connect power meter at the receiver location, facing the optical fiber, and measure the optical power of the distant lightwave equipment source. The optical power level should be the source level plus link attenuation.

7. If laser light sources are used in the link, test all optical fibers for return loss.

8. For intermittent errors or poor data throughput, connect the lightwave system to a BERT and perform a complete link. The time of errors should be logged to provide clues for cyclic problems. BERT results can be compared to the manufacturer's records. High BER can indicate high optical loss, defective lightwave equipment, or a poor electrical data connection.

If a fiber break or anomaly is found, determine its location using an OTDR. Follow the OTDR test procedures described in Chap. 13.

If fiber optic test equipment is not available, a quick fiber continuity test can be performed using a flashlight as follows:

1. Turn off and disconnect all fiber lightwave equipment.

2. Using a flashlight, shine the light into the fiber to be tested.

3. At the other end of the fiber, the light from the flashlight should be visible. The ambient room lighting may need to be dimmed in order to see the fiber light.

4. If the light from the flashlight is not visible, the fiber may be broken. Further fiber optic testing using OTDR and power meter is necessary.

This flashlight test does not provide any information on the fiber's attenuation.

22

Design Fundamentals

Designing a fiber optic system can be a complicated process. The designer must consider many factors including data rate, link attenuation, environment, cable types, fiber types, available equipment, electrical interface types, optical connectors, splicing, protocol, and so on. The complete process is quite involved and therefore beyond the scope of this book.

The process can be simplified, however, when the lightwave equipment manufacturer's instructions are followed when installing their equipment. These instructions usually provide sufficient information to select the proper optical fiber type for a simple installation. Other design considerations such as cable type, panels, jumpers, environment, routing, and the like are usually left to the designer to determine.

This chapter will show how to proceed with a simple fiber optic system design using the manufacturer's recommendations. A more involved calculation technique for selecting an optical fiber type is also discussed.

22.1 Single Mode or Multimode Fiber

The first decision that must be made is whether to install a single mode or multimode fiber optic system. Both systems have their merits.

Advantages of single mode fiber optic system

1. Single mode optical fiber has the highest possible bandwidth transmission capability and is ideal for long distance transmission.

2. Single mode optical fiber has lower attenuation than multimode fiber.

3. Single mode fiber optic cable is less expensive than multimode cable.

4. Single mode fiber is available for optical wavelengths of 1310 and 1550 nm.

Advantages of multimode fiber optic system

1. Multimode fiber is better suited for communication distances under 2 km.

2. Multimode fiber system bandwidth is more dependent on fiber length. For lengths of up to 2 km using standard 62.5/125 FDDI fiber, data rates of 100 Mbps are possible.

3. Multimode lightwave equipment is usually less expensive than single mode equipment. Inexpensive LEDs are often used as light sources.

4. Multimode fiber optic cable is usually more expensive than single mode fiber optic cable, but for short distance applications the savings in lightwave equipment may offset this cost.

5. The 62.5/125 multimode optical fiber is the standard for LAN communications such as Ethernet, token ring, and FDDI.

6. Multimode fiber is available for optical wavelengths of 850 and 1310 nm.

It can be generally concluded that single mode fiber systems are used in long distance (over 2 km) communication systems. Multimode fiber can be deployed for shorter distances as the lightwave equipment manufacturer specifies.

22.2 Basic Optical Fiber Systems

For many fiber optic installations, lightwave equipment manufacturers provide sufficient details for the user to easily implement a basic point-to-point multimode fiber optic link using their equipment. This section will present several examples of these types of installations.

Two important factors to consider in designing fiber optic links are *total link loss* and *maximum link bandwidth*. The maximum link bandwidth is generally the maximum data rate or analog bandwidth that an optical communication system can support with minimal signal distortion. It is limited by lightwave equipment properties and optical fiber parameters. Fiber optic bandwidth decreases with an increase in optical fiber length. Therefore, it is important to realize a cable length that will fit the installation and work with the given lightwave equipment. More detailed bandwidth calculations are explained in the next section. These calculations can be avoided for some basic systems because the equipment manufacturer has already tested its equipment using commercially available optical fiber for a predetermined basic system design. The manufacturer can recommend a tested optical fiber type or types to be used with its equipment for basic system installations.

Total link loss is the total light power lost in a fiber optic link due to all factors including connectors, splices, fiber attenuation, cable bends, and the like. Optical power loss due to connectors that are mounted on lightwave equipment can be ignored because they have already been included in the manufacturer's calculations. Total link loss must be within the lightwave equipment manufacturer's specifications in order for the link to function properly. This is determined by carefully planning an optical link budget, as discussed in Sec. 12.6, for the entire fiber optic system. All factors that contribute or may contribute to optical power loss are included in the link budget.

Lightwave equipment manufacturers will usually recommend an optical fiber type or a number of different optical fiber types that can be used successfully with their equipment. These optical fiber types were tested with their equipment in a standard point-to-point configuration for the indicated fiber's maximum length and maximum loss. Equipment will usually operate successfully if the recommended fiber types are implemented within the fiber loss and length restrictions. These restrictions may be listed by the equipment manufacturer in tabular form, as shown in the following example:

Fiber size (μm)	Fiber attenuation (dB/km)	Fiber NA	Fiber bandwidth (MHz × km)	Maximum loss (dB at 850 nm)	Maximum length (km)
50/125	3.0	0.20	50	2.0	0.6
50/125	2.7	0.20	50	2.0	0.7
62.5/125	3.5	0.29	50	5.0	1.4
62.5/125	3.0	0.29	50	5.0	1.6
100/140	5.0	0.29	50	9.5	1.5
100/140	4.0	0.29	50	9.5	1.8

The first three columns of this table list optical fiber specifications for a number of commercially available optical fibers. One of these fiber types should match the installed fiber's specification.

The fiber bandwidth value is the fiber's normalized bandwidth (to 1 km) and is the lowest optical fiber bandwidth that the lightwave equipment manufacturer recommends for the installation. Optical fibers with larger bandwidths are acceptable for the installation (such as 100 MHz × km in this example).

The maximum loss and maximum length should not be exceeded for the selected fiber type. The maximum loss should always be greater than or equal to the total link loss.

The maximum length is the total optical fiber length between the terminating lightwave equipment. The length represents the limit for loss due to fiber attenuation and for bandwidth due to fiber dispersion. It should not be exceeded even if the total calculated optical link loss is below the maximum loss. As this table shows, cable length increases with the size of the core diameter. This is due to the increase of the light power coupling from an LED light source into a fiber as a result of the fiber's larger diameter core and greater numerical aperture.

If the user has a choice, he or she should select a standard fiber type for an installation.

One method used to determine total link loss is the optical link budget. The optical link budget (see Sec. 12.6) lists all factors that contribute or will contribute to the system's optical loss. The result provides the required total link loss for the fiber optic system. This is then compared to the equipment's maximum loss to determine whether the design is within attenuation specification.

Procedure

1. Obtain the following information from the lightwave equipment manufacturer.

   - optical fiber diameter recommendations, 8/125, 50/125, 62.5/125, 100/140
   - maximum optical fiber attenuation dB/km recommendation
   - optical fiber numerical aperture (NA) recommendation
   - maximum optical fiber bandwidth (MHz × km) at operating wavelength recommendation
   - optical fiber maximum length recommendation
   - equipment's maximum loss specification
   - equipment's receiver sensitivity at BER
   - equipment transmitter average output power
   - equipment receiver dynamic range

   If the maximum loss is provided and the receiver has a full dynamic range (it will operate with full and minimum light power), the specifications for the receiver's sensitivity and transmitter's average output power are not necessary.

Maximum loss = transmitter's average output power − receiver's sensitivity

2. From the fiber optic installation plan determine

- the total fiber optic link length
- the number of required optical splices and the loss per splice
- the number of fiber connections and the loss per connection
- a design margin
- optical losses due to any other components in the system

3. Complete the optical budget as described in Sec. 12.6:

- Optical fiber loss at operating wavelength: kilometer length at dB/km
- Splice loss: splices at dB/splice
- Connection loss: connections at dB/connection
- Other component losses
- Design margin
- Total link loss
- Transmitter's average output power
- Receiver's input power
- Receiver's dynamic range
- Receiver's sensitivity at BER
- Remaining margin

4. The remaining margin should be greater than zero for proper system design. If it is not, reexamine all loss values to reduce total link loss.

Selecting optical fiber: An example   A fiber optic link should be designed to provide point-to-point data communications between two computers (see Fig. 22.1). Lightwave communication equipment that is compatible with the computer equipment (proper electrical interface and communication protocol) has been selected. The equipment manufacturer's recommended optical fiber specification is presented as a table later in this example. In addition, the receiver's dynamic range is from zero to full transmitter output 0 to 10 dB.

The cable length through an outdoor route has been measured and is 1.2 km. Due to the nature of the installation, four separate fiber cables will be required to complete the link. A patch panel is requested at both ends for easy connection of patch cords. What type of optical fiber should be selected for the installation?

Fiber size (μm)	Fiber attenuation (dB/km)	Fiber NA	Fiber bandwidth (MHz × km)	Maximum loss (dB at 850 nm)	Maximum length (km)
50/125	3.0	0.20	50	2.0	0.6
50/125	2.7	0.20	50	2.0	0.7
62.5/125	3.5	0.29	50	5.0	1.4
62.5/125	3.0	0.29	50	5.0	1.6
100/140	5.0	0.29	50	9.5	1.5
100/140	4.0	0.29	50	9.5	1.8

The two major considerations for optical fiber selection are total link loss and optical fiber length. The total length between lightwave terminating equipment has been measured and is 1.2 km. It should be noted that any future extension to the fiber link should be considered here. Both optical link loss and fiber cable length will increase if an extension fiber is added to accommodate additional future requirements. If not properly planned, an extension to the fiber cable may not be possible. For this example, future extension of the link is not required.

The first step is to compile all known information to determine an optical link budget for the installation.

The fiber optic cable length is 1.2 km. Three splices are required to connect the four sections of cable together. Because the cables will be spliced outdoors, the fusion splicing method has been selected, with a maximum loss of 0.1 dB per splice. Two patch panels are to be used with patch cords connecting the equipment. The fiber optic cable is to be directly terminated with connectors. Connection loss at each patch panel is 1.0 dB. Patch cord connection loss at equipment is not added to the link budget because it is already factored into the manufacturer's specifications. Patch cords are quite short at 3 m, so their fiber attenuation is minimal and is ignored.

1. The following information has been obtained from the lightwave equipment and cable manufacturers:

   ■ recommended optical fiber type and diameter(s), in table form for six optical fibers

   ■ recommended optical fiber maximum attenuation (dB/km) at operating wavelength, in table form for six optical fibers

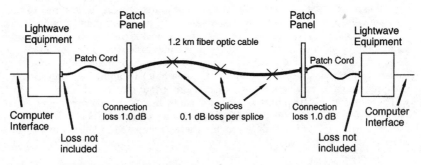

Figure 22.1   Fiber optic computer link example.

- recommended optical fiber numerical aperture (NA), in table form for six optical fibers
- recommended optical fiber bandwidth (MHz × km) at operating wavelength, in table form for six optical fibers
- optical fiber maximum length, in table form for six optical fibers
- equipment maximum loss specification for optical fiber type used, in table form for six optical fibers
- equipment receiver sensitivity at BER not provided
- equipment transmitter average output power not provided
- equipment receiver dynamic range, from full source power to minimum receiver sensitivity (a full dynamic range)

2. From the fiber optic installation plan:

- the total fiber optic link length is 1.2 km
- the number of required optical splices is three at 0.1 dB per splice
- the number of fiber connections is two at 1 dB per connection
- a design margin estimate at 2 dB
- optical losses due to any other components in the system is zero

From the lightwave equipment manufacturer's recommended optical fiber list, the distance criteria can be met with any of the 62.5/125 or 100/140 optical fibers. The first choice for an optical fiber is a standard 62.5/125 with 3.0 dB/km loss. This is used for a preliminary optical budget calculation as follows.

3. Optical budget:

Optical fiber loss at 850 nm:	
1.2-km length at 3.0 dB/km	3.6 dB
Splice loss:	
Three splices at 0.1 dB/splice	0.3 dB
Connection loss:	
Two connections at 1.0 dB/connection	2.0 dB
Other components' loss	0
Optical margin	2.0 dB
Total link loss	7.9 dB

Therefore, using the 62.5/125 μm, 3.0-dB/km optical fiber will result in a total fiber link loss of 7.9 dB. This is higher than the lightwave equipment manufacturer's maximum loss of 5 dB and therefore cannot be used.

A second choice will be the 100/140 fiber with 4 dB/km attenuation. The link budget then looks as follows:

Optical fiber loss at 850 nm:	
1.2-km length at 4.0 dB/km	4.8 dB
Splice loss:	
Three splices at 0.1 dB/splice	0.3 dB
Connection loss:	
Two connections at 1.0 dB/connection	2.0 dB
Other components' loss	0
Optical margin	2.0 dB
Total link loss	9.1 dB

The total link loss using the 100/140-μm at 4 dB/km fiber is 9.1 dB. This is less than the lightwave equipment manufacturer's maximum loss of 9.5 dB. Therefore, the maximum loss criterion has been satisfied. The lightwave equipment manufacturer's maximum length of 1.8 km is above the required 1.2-km installation length, and therefore the length criterion has also been satisfied. Consequently, this optical fiber type can be used for the installation.

Many LAN lightwave equipment manufacturers will recommend that one standard optical fiber (62.5/125) be used with their equipment. This simplifies the procedure.

The following example shows calculations based on this information.

Standard optical fiber used with LAN lightwave equipment: An example A LAN segment in a plant is to be extended using fiber optics. Two LAN optical fiber repeaters designed specifically for this purpose are being considered for the link. The repeater manufacturer provides the following information for the optical fiber to be used with the equipment:

Equipment operating wavelength	850 nm
Optical fiber type	multimode 62.5/125, NA = 0.29
Optical fiber bandwidth	100 MHz × km
Maximum fiber attenuation	5.0 dB/km
Maximum optical fiber length	1 km
Receiver dynamic range	full range

The equipment manufacturer assumes that the optical fiber will not be spliced, that only two connectors at the equipment will be required, and that no other attenuation will be added to the link.

The length of the fiber optic link is measured to be 0.7 km. The fiber optic cable will be dedicated and will not require patch panels or splices. What optical fiber should be purchased for the link to operate? What will the design look like?

This design is simple, and the equipment manufacturer provides all the information required for proper optical fiber purchase.

The measured optical fiber length will be 0.7 km, which is below the manufacturer's maximum length criteria. No splices or connectors (other than directly terminating cable connectors) are used. The only loss will result from optical fiber attenuation. The cable to select should provide the same or better specification than the manufacturer's recommendation. The design will resemble that shown in Fig. 22.2, a dedicated fiber optic cable with no intermediate splice or connections.

If splices or connectors are to be added, then a link budget should be determined. For example, if two connectors and one splice are added to the fiber link, then the budget will look as follows:

Optical fiber loss at 850 nm:	
0.7 km length at ? dB/km	? dB
Splice loss:	
One splice at 0.1 dB/splice	0.1 dB
Connection loss:	
Two connections at 1.0 dB/connection	2.0 dB
Other components' loss	0 dB
Optical margin	2.0 dB
Total link loss	5.0 dB

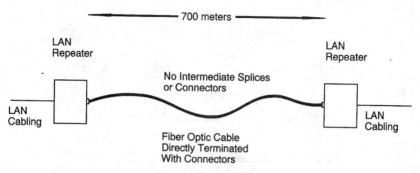

Figure 22.2   Simple fiber optic LAN.

The equipment manufacturer's maximum fiber attenuation is converted to a maximum loss by multiplying by the 1-km maximum length. The resultant 5 dB is used as a total link loss. Rearranging the equation, optical fiber loss is calculated to be 1.9 dB, and the optical fiber attenuation should be 2.7 dB/km or less.

$$\text{Optical fiber loss} = 5.0 - 2.0 - 2.0 - 0.1$$

$$\text{Optical fiber loss} = 0.9 \text{ dB}$$

$$\text{Optical fiber attenuation} = 0.9 \text{ dB}/0.7 \text{ km}$$

$$\text{Maximum optical fiber attenuation} = 1.2 \text{ dB/km}$$

A maximum attenuation of 1.2-dB/km optical fiber is required if two connectors and a splice are to be added to the link.

In the next example, the equipment manufacturer provides nominal optical output power and receiver sensitivity instead of attenuation or loss level. This requires additional calculations but is not complicated.

Nominal optical output power and receiver sensitivity: An example   A fiber optic link is to be used to connect a remote video surveillance camera to a monitor that is 3 km away (see Fig. 22.3). The camera is a high-quality model with a 10 MHz bandwidth and standard NTSC video output. A manufacturer of lightwave equipment that can convert the 10 MHz NTSC electrical signal to an optical transmission is found, and it has the following equipment specifications:

Equipment operating wavelength	1310 nm
Optical fiber type	multimode 62.5/125, NA = 0.29
Optical fiber bandwidth	300 MHz × km
Nominal optical output power	−15 dBm
Receiver sensitivity	−25 dBm at S/N 68 dB
Maximum optical fiber length	3 km
Receiver dynamic range	−20 dBm to −25 dBm

Two patch panels and two splices will be required for the installation. Can this lightwave equipment be used for this installation? If so, what optical fiber is required?

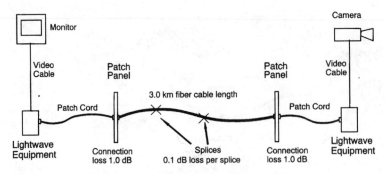

Figure 22.3 Fiber optic video link example.

The first step is to determine an optical link budget for the installation. The fiber optic cable length is 3 km. Two splices are required to connect the three sections of cable together. Because the cables will be spliced outdoors, the fusion splicing method has been selected with a loss of 0.1. The cable will be directly terminated with connectors. Two patch panels are to be used with patch cords to connect the equipment. Connection loss at each patch panel is 1.0 dB. Patch cord connection loss at the equipment is not added to the link budget because it is already factored into the manufacturer's specifications. Patch cords are quite short at 3 m, so their fiber attenuation is minimal and is ignored.

1. The following information was obtained from the equipment manufacturer:
   - optical fiber diameter recommendation 62.5/125
   - optical fiber attenuation to be determined
   - optical fiber numerical aperture NA 0.29
   - optical fiber bandwidth at operating wavelength 300 MHz × km
   - optical fiber maximum length 3 km
   - equipment maximum loss not provided
   - equipment receiver sensitivity -25 dBm at 68 S/N
   - equipment transmitter average output power -15 dBm
   - equipment receiver dynamic range is -20 to -25 dBm

2. From the fiber optic installation plan:
   - the total fiber optic link length is 3 km
   - the number of required optical splices is two at 0.1 dB per splice
   - the number of fiber connections is two at 1 dB per connection
   - a design margin estimate of 2 dB
   - optical losses due to any other components in the system is zero

3. The optical budget calculations is as follows:

Optical fiber loss at 1310 nm:	
3.0-km length at ? dB/km	? dB
Splice loss:	
Two splices at 0.1 dB/splice	0.2 dB

Connection loss:

Two connections at 1.0 dB/connection	2.0 dB
Other components' loss	0 dB
Optical margin	2.0 dB
Total link loss	? dB
Transmitter's average output power	− 15.0 dBm
Receiver's input power (g − f)	? dBm
Receiver's dynamic range	−20 to −25 dBm
Receiver's sensitivity at 68 dB S/N	−25 dBm
Remaining margin (h − j)	0 dB

Working backward, we solve for receiver input power using a remaining margin of 0 dB and receiver sensitivity of −25 dBm.

$$\text{Receiver input power} = \text{Remaining margin} + \text{receiver sensitivity}$$

$$\text{Receiver input power} = 0 \text{ dB} + (-25 \text{ dBm})$$

$$\text{Receiver input power} = -25 \text{ dBm}$$

Next we solve for total link loss:

$$\text{Total link loss} = \text{Transmitter average output power} - \text{receiver input power}$$

$$\text{Total link loss} = -15 \text{ dBm} - (-25 \text{ dBm})$$

$$\text{Total link loss} = 10 \text{ dB}$$

To determine the optical fiber loss use the following formula:

$$\text{Optical fiber loss} = \text{Total link loss} - \text{Connection loss}$$
$$- \text{Optical margin} - \text{Splice loss}$$

$$\text{Optical fiber loss} = 10.0 - 2.0 - 2.0 - 0.2$$

$$\text{Optical fiber loss} = 5.8 \text{ dB}$$

The optical fiber loss is divided by the total cable length to determine the optical fiber attenuation:

$$\text{Optical fiber attenuation} = 5.8 \text{ dB}/3.0 \text{ km}$$

$$\text{Optical fiber attenuation} = 1.9 \text{ dB/km}$$

The 62.5/125 optical fiber (NA = 0.29) to be used for this installation should have an attenuation no greater than 1.9 dB/km at 1310 nm. To satisfy the bandwidth criterion, the cable should have a bandwidth of 300 MHz × km and be shorter than 3 km.

In these examples, a 2-dB optical design margin was used to account for any future contingencies such as additional fiber splices, dirty connections, or cable extension.

22.3 Bandwidth Calculations

For some fiber optic installations it may be necessary to determine the maximum bandwidth that a fiber optic system can successfully support. This may be required because the lightwave equipment manufacturer did not provide sufficient details for the installation or because the system design is complex. The best method to determine maximum system bandwidth is to directly measure it. However, this is time-consuming and requires elaborate equipment. Theoretical equations can also be used to predict the system bandwidth, but this involves complex calculations that are beyond the scope of this book. A useful approximation can be made using the method described in this section.

Multimode fiber

The multimode fiber system transmission bandwidth is limited by the optical fiber parameters, modal dispersion, and chromatic dispersion and also by the lightwave equipment parameters, light source, and photodetector rise times. These parameters should be considered when making the bandwidth calculations.

In order to approximate a multimode fiber's maximum data rate, accurate fiber and equipment data must be obtained from the manufacturer. The information required is as follows:

Fiber data

Optical fiber modal bandwidth (at operating wavelength):	MHz × km	$B_{\text{Modal(MHz} \times \text{km)}}$
Fiber chromatic dispersion (at operating wavelength):	ns/nm × km	$D_{\text{Chromatic(ns/nm} \times \text{km)}}$
Total fiber installation length:	km	$d_{\text{Install(km)}}$
Optical fiber factory length:	km	$d_{\text{Factory(km)}}$
Gamma, cutback/concatenation:	gamma	γ

Lightwave equipment data

Transmission baud rate:	Mbps	$R_{\text{(Mbps)}}$
Optical modulation method:	NRZ, RZ, Manchester, Analog	
Light source spectral width:	nm	$W_{\text{Spectral (nm)}}$
Light source rise time:	ns	$T_{\text{LightSource (ns)}}$
Photo detector rise time:	ns	$T_{\text{Detector (ns)}}$
Operating wavelength:	nm	λ

For multimode fiber, both chromatic dispersion and multimode (modal) dispersion must be considered when determining a fiber optic link's maximum data transmission rate. Manufacturers' specifications list the optical fiber's modal bandwidth as a function of distance as

MHz × km. This value does not include chromatic dispersion, which is provided by the manufacturer as a fiber rise time in the unit ns/nm × km. Both these units are required to determine a final bandwidth value.

An accurate optical fiber length for the installation must be determined for these calculations. This length is the total fiber installation length between the lightwave transmitter and the receiver's terminating equipment.

The fiber factory length is the fiber length that the manufacturer has used to measure the optical fiber modal bandwidth (MHz × km).

The cutback gamma is the gamma value provided by the manufacturer for installed fiber lengths shorter than the factory-measured optical fiber modal bandwidth length. The concatenation gamma is the gamma value provided by the manufacturer for installed fiber lengths that are longer than the factory-measured optical fiber modal bandwidth length.

The transmission baud rate is the equipment's electrical *baud rate*. The modulation method is commonly a PCM type with NRZ, RZ, Manchester coding, or analog amplitude modulation.

The light source's spectral width is the half optical power spectral width (−3 dB) in nanometers of the light source.

The source and detector rise time is the time in nanoseconds required for an input step waveform to rise between 10 and 90 percent of the total amplitude at the output.

The operating wavelength is the optical wavelength that the system will be used at for communications. All testing should be performed, and fiber parameters should be determined for this wavelength.

The following steps can be followed to determine the optical fiber type given the lightwave equipment's parameters.

1. First, the system electrical bandwidth ($B_{\text{ElectricalSystem[MHz]}}$) should be determined given the system's required data rate in Mbps. The bandwidth depends on the modulation method. An NRZ code modulation requires half as much bandwidth as a RZ code modulation:

$$B_{\text{ElectricalSystem(MHz)}} = R_{\text{RZ(Mbps)}}$$

$$B_{\text{ElectricalSystem(MHz)}} = R_{\text{Manchester(Mbps)}}$$

$$B_{\text{ElectricalSystem(MHz)}} = F_{\text{Analog(MHz)}}$$

$$B_{\text{ElectricalSystem(MHz)}} = \frac{R_{\text{NRZ(Mbps)}}}{2}$$

2. A rise-time budget equation will be applied to solve for the unknown parameters:

$$T^2_{\text{Modal(ns)}} + T^2_{\text{Chromatic(ns)}} = T^2_{\text{System(ns)}} - T^2_{\text{LightSource(ns)}} - T^2_{\text{Detector(ns)}}$$

The left-hand side of this equation can be referred to as the total optical fiber rise time.

$$T^2_{\text{Fiber(ns)}} = T^2_{\text{Modal(ns)}} + T^2_{\text{Chromatic(ns)}}$$

The right-hand side of the equation can be referred to as the equipment rise time:

$$T^2_{\text{Equipment}} = T^2_{\text{System(ns)}} - T^2_{\text{LightSource(ns)}} - T^2_{\text{Detector(ns)}}$$

3. Rise time attributed to equipment parameters is calculated first. The electrical system bandwidth is converted into a rise time using a conservative approximation:

$$T_{\text{System(ns)}} = 0.35/B_{\text{ElectricalSystem(GHz)}}$$

4. Using the manufacturer's data, the light source and detector rise times are entered into the equipment rise time equation, which is then solved:

$$T^2_{\text{Equipment}} = T^2_{\text{System(ns)}} - T^2_{\text{LightSource(ns)}} - T^2_{\text{Detector(ns)}}$$

5. Next, the values for the fiber's total rise time are determined. The chromatic dispersion rise time is calculated:

$$T_{\text{Chromatic(ns)}} = D_{\text{Chromatic(ns/nm} \times \text{km)}} \times W_{\text{Spectral(nm)}} \times d_{\text{Install(km)}}$$

6. The optical fiber modal bandwidth length product is converted into a total installed optical fiber modal bandwidth for the entire fiber length:

$$B_{\text{ModalInstall(MHz)}} = B_{\text{Modal(MHz} \times \text{km)}}/d_{\text{Install(km)}}$$

If the cable length when installed is not the same as the factory length, then the manufacturer's gamma factor should be applied to adjust the bandwidth to the installed length.

Optical fiber modal bandwidth values are measured at the factory for a standard manufactured length. If this fiber length is decreased or increased for an installation, the fiber modal bandwidth will change. The gamma factor provides a correction for this change in length. The optical fiber's factory-measured modal bandwidth length can usually be obtained from the manufacturer. The gamma (γ) value has no units and varies from 0.5 to 1.0.

If the optic fiber's installation length is less than the factory-measured length (fiber will be cut back), the following equation can be applied:

$$B_{\text{ModalInstall(MHz)}} = (B_{\text{Modal(MHz} \times \text{km)}}/d_{\text{Factory(km)}}) \times (d_{\text{Factory(km)}}/d_{\text{Install(km)}})^{\gamma}$$

If the optical fiber's installation length is greater than the factory-measured length, a number of factory fiber lengths will be concatenated, and the following equation can be applied:

$$B_{\text{ModalInstall(MHz)}} = \left[\sum\nolimits_{\text{NumberOfFibers}} \left(B_{\text{Modal(MHz} \times \text{km)}}/d_{\text{Factory(km)}} \right)^{-1/\gamma} \right]^{-\gamma}$$

7. The optical fiber's modal bandwidth is converted into an electrical modal bandwidth. This is required because a 3-dB drop in optical power is equivalent to a 6-dB drop in electrical power.

$$B_{\text{ModalElectrical(MHz)}} = 0.71 \times B_{\text{ModalInstall(MHz)}}$$

8. The electrical modal bandwidth is then converted to a modal rise time:

$$T_{\text{Modal(ns)}} = 0.35/B_{\text{ModalElectrical(GHz)}}$$

9. The total fiber rise time is then calculated:

$$T^2_{\text{Fiber(ns)}} = T^2_{\text{Modal(ns)}} + T^2_{\text{Chromatic(ns)}}$$

The optical fiber's total electrical bandwidth can then be calculated as follows:

$$B_{\text{FiberTotal(GHz)}} = 0.35/T_{\text{Fiber(ns)}}$$

10. If the total fiber rise time $T_{\text{Fiber(ns)}}$ is equal to or less than the total equipment rise time $T_{\text{Equipment}}$, the optical fiber selection is appropriate for the required equipment data transmission rate.

Data LAN communication link: An example   A fiber optic link is to be designed in order to provide data LAN communication at 10 Mbps. The distance of the fiber optic route between LAN repeaters is 3.2 km. The fiber optic cable is purchased at a factory length of 4.4 km and will then be cut back for the installation. The lightwave equipment to be selected has the following data specifications:

Maximum baud rate:	10 Mbps
Modulation method:	NRZ
Lightwave spectral width:	20 nm
Light source rise time:	12 ns
Photo detector rise time:	20 ns
Transmission wavelength:	850 nm

The following fiber type is contemplated for the installation:

Fiber type:	Multimode 62.5/125 μm
Bandwidth at 850 nm:	160 MHz × km
Chromatic dispersion:	0.1 ns/nm × km
Factory length:	4.4 km
Installation length:	3.2 km
Cutback gamma:	0.5

Will this optical fiber support the lightwave equipment transmission data rate?

1. First, the system's electrical bandwidth is determined:

$$B_{\text{ElectricalSystem(MHz)}} = R_{\text{NRZ(Mbps)}}/2$$
$$B_{\text{ElectricalSystem(MHz)}} = 10/2$$
$$B_{\text{ElectricalSystem(MHz)}} = 5 \text{ MHz}$$

2. A rise-time budget equation will then be applied to determine if the fiber's bandwidth is acceptable:

$$T^2_{\text{Modal(ns)}} + T^2_{\text{Chromatic(ns)}} = T^2_{\text{System(ns)}} - T^2_{\text{LightSource(ns)}} - T^2_{\text{Detector(ns)}}$$

3. The electrical system bandwidth is converted into an approximate rise time:

$$T_{\text{System(ns)}} = 0.35/B_{\text{ElectricalSystem(GHz)}}$$
$$T_{\text{System(ns)}} = 0.35/0.005$$
$$T_{\text{System(ns)}} = 70 \text{ ns}$$

4. Rise time attributed to lightwave equipment parameters can be calculated:

$$T^2_{\text{Equipment}} = T^2_{\text{System(ns)}} - T^2_{\text{LightSource(ns)}} - T^2_{\text{Detector(ns)}}$$
$$T^2_{\text{Equipment}} = 70^2 - 12^2 - 20^2$$
$$T_{\text{Equipment}} = 66 \text{ ns}$$

5. The optical fiber's chromatic dispersion rise time is calculated:

$$T_{\text{Chromatic(ns)}} = D_{\text{Chromatic(ns/nm × km)}} \times W_{\text{Spectral(nm)}} \times d_{\text{Install(km)}}$$
$$T_{\text{Chromatic(ns)}} = 0.1 \times 20 \times 3.2$$
$$T_{\text{Chromatic(ns)}} = 6.4 \text{ ns}$$

6. The optical fiber's installation length is less than the factory-measured bandwidth length; therefore, the gamma cutback equation will be used to adjust the fiber modal bandwidth value:

$$B_{\text{ModalInstall(MHz)}} = (B_{\text{Modal(MHz × km)}}/d_{\text{Factory(km)}}) \times (d_{\text{Factory(km)}}/d_{\text{Install(km)}})^{\gamma}$$
$$B_{\text{ModalInstall(MHz)}} = 160/4.4 \times (4.4/3.2)^{0.5}$$
$$B_{\text{ModalInstall(MHz)}} = 42.6 \text{ MHz}$$

7. The optical modal bandwidth is converted into an electrical bandwidth:

$$B_{\text{ModalElectrical(MHz)}} = 0.71 \times B_{\text{ModalInstall(MHz)}}$$
$$B_{\text{ModalElectrical(MHz)}} = 0.71 \times 42.6$$
$$B_{\text{ModalElectrical(MHz)}} = 30.2 \text{ MHz}$$

8. The electrical modal bandwidth is then converted to a modal rise time:

$$T_{\text{Modal(ns)}} = 0.35/B_{\text{ModalElectrical(GHz)}}$$

$$T_{\text{Modal(ns)}} = 0.35/0.0302$$

$$T_{\text{Modal(ns)}} = 11.6 \text{ ns}$$

9. Therefore, the total fiber rise time is as follows:

$$T^2_{\text{Fiber(ns)}} = T^2_{\text{Modal(ns)}} + T^2_{\text{Chromatic(ns)}}$$

$$T^2_{\text{Fiber(ns)}} = 11.6^2 + 6.4^2$$

$$T_{\text{Fiber(ns)}} = 13.2 \text{ ns}$$

The optical fiber total electrical bandwidth is as follows:

$$B_{\text{FiberTotal(GHz)}} = 0.35/T_{\text{Fiber(ns)}}$$

$$B_{\text{FiberTotal(GHz)}} = 0.35/13.2$$

$$B_{\text{FiberTotal(MHz)}} = 26.5 \text{ MHz}$$

10. The total optical fiber rise time is 13.2 ns, which is less than the equipment's required rise time of 66 ns. Therefore, this optical fiber selection is appropriate for the required data transmission rate.

Closed circuit video transmission link: An example A fiber optic link is to be designed to provide a video transmission link for a closed circuit monitoring video camera. The distance of the fiber optic route between optical modems is 4.4 km. The optical modem equipment to be selected has the following data specifications:

Video 3 dB bandwidth:	10 MHz
Modulation method:	analog
Lightwave spectral width:	20 nm
Light source rise time:	5 ns
Photo detector rise time:	8 ns
Transmission wavelength:	850 nm

The following fiber type is contemplated for the installation:

Fiber type:	Multimode 62.5/125 μm
Bandwidth at 1310 nm:	200 MHz × km
Chromatic dispersion:	0.1 ns/nm × km
Factory length:	2.2 km
Installation length:	4.4 km
Concatenation gamma:	0.9

Will this optical fiber have sufficient bandwidth for the required cable length?

1. First, the system's electrical bandwidth is determined:

$$B_{\text{ElectricalSystem(MHz)}} = R_{\text{Analog(MHz)}}$$
$$B_{\text{ElectricalSystem(MHz)}} = 10 \text{ MHz}$$

2. A rise time budget equation will be applied to determine if the fiber's bandwidth is acceptable:

$$T^2_{\text{Modal(ns)}} + T^2_{\text{Chromatic(ns)}} = T^2_{\text{System(ns)}} - T^2_{\text{LightSource(ns)}} - T^2_{\text{Detector(ns)}}$$

3. The electrical system bandwidth is converted into an approximate rise time:

$$T_{\text{System(ns)}} = 0.35/B_{\text{ElectricalSystem(GHz)}}$$
$$T_{\text{System(ns)}} = 0.35/0.01$$
$$T_{\text{System(ns)}} = 35 \text{ ns}$$

4. Rise time attributed to lightwave equipment parameters can be calculated as follows:

$$T^2_{\text{Equipment}} = T^2_{\text{System(ns)}} - T^2_{\text{LightSource(ns)}} - T^2_{\text{Detector(ns)}}$$
$$T^2_{\text{Equipment}} = 35^2 - 5^2 - 8^2$$
$$T_{\text{Equipment}} = 33.7 \text{ ns}$$

5. The optical fiber chromatic dispersion rise time is then calculated:

$$T_{\text{Chromatic(ns)}} = D_{\text{Chromatic(ns/nm} \times \text{km)}} \times W_{\text{Spectral(nm)}} \times d_{\text{(km)}}$$
$$T_{\text{Chromatic(ns)}} = 0.1 \times 20 \times 4.4$$
$$T_{\text{Chromatic(ns)}} = 8.8 \text{ ns}$$

6. The optical fiber's installation length requires two factory cable lengths; therefore, the gamma concatenation equation will be used to adjust the fiber modal bandwidth value.

$$B_{\text{ModalInstall(MHz)}} = \left[\Sigma_{\text{NumberOfFibers}} (B_{\text{Modal(MHz} \times \text{km)}}/d_{\text{Factory(km)}})^{-1/\gamma} \right]^{-\gamma}$$
$$B_{\text{ModalInstall(MHz)}} = [(200/2.2)^{-1/0.9} + (200/2.2)^{-1/0.9}]^{-0.9}$$
$$B_{\text{ModalInstall(MHz)}} = 48.7 \text{ MHz}$$

7. The optical modal bandwidth is then converted into an electrical bandwidth:

$$B_{\text{ModalElectrical(MHz)}} = 0.71 \times B_{\text{ModalInstall(MHz)}}$$
$$B_{\text{ModalElectrical(MHz)}} = 0.71 \times 48.7$$
$$B_{\text{ModalElectrical(MHz)}} = 34.6 \text{ MHz}$$

8. The electrical modal bandwidth is then converted to a modal rise time:

$$T_{\text{Modal(ns)}} = 0.35/B_{\text{ModalElectrical(GHz)}}$$
$$T_{\text{Modal(ns)}} = 0.35/0.0346$$
$$T_{\text{Modal(ns)}} = 10.1 \text{ ns}$$

9. Therefore, the total fiber rise time is as follows:

$$T^2_{\text{Fiber(ns)}} = T^2_{\text{Modal(ns)}} + T^2_{\text{Chromatic(ns)}}$$
$$T^2_{\text{Fiber(ns)}} = 10.1^2 + 8.8^2$$

$$T_{\text{Fiber(ns)}} = 13.4 \text{ ns}$$

The optical fiber total electrical bandwidth is as follows:

$$B_{\text{FiberTotal(GHz)}} = 0.35/T_{\text{Fiber(ns)}}$$
$$B_{\text{FiberTotal(GHz)}} = 0.35/13.4$$
$$B_{\text{FiberTotal(MHz)}} = 26.1 \text{ MHz}$$

10. The total optical fiber rise time is 13.4 ns, which is less than the equipment's required rise time of 33.7 ns. Therefore, this optical fiber selection is appropriate for the required video bandwidth.

Single mode fiber

Single mode fiber system bandwidth is limited by the fiber's chromatic material dispersion and the chromatic waveguide dispersion, which is specified in the form, picoseconds/(nanometer × kilometer), (ps/nm × km). It is also limited by equipment parameters such as light source and photodetector rise times. Conventional single mode fibers are available with near-zero dispersion at 1310-nm operating wavelength, thereby supporting very high bandwidths. Optical fibers with near-zero dispersion at 1550 nm are available and are known as *dispersion-shifted fibers*. Optical fibers with near-zero dispersion at both 1310 and 1550 nm are also available and are known as *dispersion-flattened fibers*.

In order to approximate a single mode fiber's maximum data rate, accurate fiber and equipment data must be obtained from the manufacturer. The following information is required:

Fiber data

Chromatic material dispersion (at operating wavelength):	ns/nm × km	$D_{\text{Material(ns/nm × km)}}$
Chromatic waveguide dispersion (at operating wavelength):	ns/nm × km	$D_{\text{Waveguide(ns/nm × km)}}$
Total fiber installation length:	km	$d_{\text{Install(km)}}$

Lightwave equipment data

Transmission baud rate:	Mbps	$R_{\text{(Mbps)}}$
Optical modulation method:	NRZ, RZ, Manchester, Analog	
Light source spectral width:	nm	$W_{\text{Spectral(nm)}}$
Light source rise time:	ns	$T_{\text{LightSource(ns)}}$
Photo detector rise time:	ns	$T_{\text{Detector(ns)}}$
Operating wavelength(s):	nm	λ

Both material and waveguide dispersion must be considered to determine a fiber optic bandwidth. Manufacturers' specifications list the dispersion in the units ps/nm × km.

An accurate installation length for the optical fiber must be determined for these calculations. This length is the total optical fiber link length between lightwave transmitter and receiver terminating equipment.

The transmission baud rate is the equipment's electrical baud rate in Mbps. The modulation method is commonly a PCM type with NRZ, RZ, Manchester coding, or analog amplitude modulation.

The light source spectral width is the half optical power spectral width (-3 dB) in nanometers of the light source and is often a laser.

The source and detector rise time is the time in nanoseconds required for an input step waveform to rise between 10 and 90 percent of total amplitude at the output.

The following steps can be followed to determine the optical fiber type given lightwave equipment parameters:

1. First, the system electrical bandwidth ($B_{\text{ElectricalSystem(MHz)}}$) should be determined given the system's required data rate in Mbps. The bandwidth depends on the modulation method. An NRZ code modulation requires half as much bandwidth as an RZ code modulation:

$$B_{\text{ElectricalSystem(MHz)}} = R_{\text{RZ(Mbps)}}$$

$$B_{\text{ElectricalSystem(MHz)}} = R_{\text{Manchester(Mbps)}}$$

$$B_{\text{ElectricalSystem(MHz)}} = F_{\text{Analog(MHz)}}$$

$$B_{\text{ElectricalSystem(MHz)}} = \frac{R_{\text{NRZ(Mbps)}}}{2}$$

2. A rise-time budget equation will be applied to solve for the unknown parameters:

$$T^2_{\text{Chromatic(ns)}} = T^2_{\text{System(ns)}} - T^2_{\text{LightSource(ns)}} - T^2_{\text{Detector(ns)}}$$

The left-hand side of this equation can be referred to as the total optical fiber rise time:

$$T_{\text{Fiber(ns)}} = T_{\text{Chromatic(ns)}}$$

The right-hand side of the equation can be referred to as the equipment rise time:

$$T^2_{\text{Equipment}} = T^2_{\text{System(ns)}} - T^2_{\text{LightSource(ns)}} - T^2_{\text{Detector(ns)}}$$

3. Rise time attributed to equipment parameters is calculated first. The electrical system bandwidth is converted into a rise time using a conservative approximation:

$$T_{\text{System(ns)}} = 0.35/B_{\text{ElectricalSystem(GHz)}}$$

4. Using the manufacturer's data, the light source and detector rise times are entered into the equipment rise-time equation, which is then solved:

$$T^2_{\text{Equipment}} = T^2_{\text{System(ns)}} - T^2_{\text{LightSource(ns)}} - T^2_{\text{Detector(ns)}}$$

5. The chromatic rise time is calculated using the fiber chromatic dispersion values at the operating wavelength:

$$D_{\text{Chromatic(ns/nm} \times \text{km)}} = D_{\text{Material(ns/nm} \times \text{km)}} + D_{\text{Waveguide(ns/nm} \times \text{km)}}$$
$$T_{\text{Chromatic(ns)}} = D_{\text{Chromatic(ns/nm} \times \text{km)}} \times W_{\text{Spectral(nm)}} \times d_{\text{Install(km)}}$$

The optical fiber's total electrical bandwidth is as follows:

$$B_{\text{FiberTotal(GHz)}} = 0.35/T_{\text{Fiber(ns)}}$$

6. If the total fiber rise time $T_{\text{Chromatic(ns)}}$ is equal to or less than the total equipment rise time $T_{\text{Equipment}}$, the optical fiber selection is appropriate for the required equipment data transmission rate.

OC-3 communication link: An example   A fiber optic link is designed to provide a high-speed OC-3 communication link to provide 2000 telephone communication channels between two cities 50 km apart. The lightwave equipment that is selected has the following specifications:

Transmission baud rate:	155.52 Mbps
Optical modulation method:	NRZ type
Light source spectral width:	20 nm
Light source rise time:	2.0 ns
Photo detector rise time:	1.0 ns
Operating wavelength:	1310 nm

The optical fiber to be selected has the following specifications:

Chromatic material dispersion:	-6 ps/nm $\times$ km at 1310 nm
Chromatic waveguide dispersion:	5 ps/nm $\times$ km at 1310 nm
Total fiber installation length:	50 km

Will this optical fiber provide the appropriate transmission bandwidth?

1. First, the system electrical bandwidth is determined:

$$B_{\text{ElectricalSystem(MHz)}} = \frac{R_{\text{NRZ(Mbps)}}}{2}$$

$$B_{\text{ElectricalSystem(MHz)}} = \frac{155.52}{2}$$

$$B_{\text{ElectricalSystem(MHz)}} = 77.76 \text{ MHz}$$

2. A rise-time budget equation will be applied to solve for the unknown parameters:

$$T^2_{\text{Chromatic(ns)}} = T^2_{\text{System(ns)}} - T^2_{\text{LightSource(ns)}} - T^2_{\text{Detector(ns)}}$$

3. The electrical system bandwidth is converted into a rise time:

$$T_{\text{System(ns)}} = 0.35/B_{\text{ElectricalSystem(GHz)}}$$
$$T_{\text{System(ns)}} = 0.35/0.07776$$
$$T_{\text{System(ns)}} = 4.5 \text{ ns}$$

4. The equipment rise-time equation is then solved:

$$T^2_{\text{Equipment}} = T^2_{\text{System(ns)}} - T^2_{\text{LightSource(ns)}} - T^2_{\text{Detector(ns)}}$$
$$T^2_{\text{Equipment}} = 4.5^2 - 2.0^2 - 1.0^2$$
$$T_{\text{Equipment}} = 3.9 \text{ ns}$$

5. The chromatic dispersion rise time is calculated:

$$D_{\text{Chromatic(ns/nm} \times \text{km)}} = D_{\text{Material(ns/nm} \times \text{km)}} + D_{\text{Waveguide(ns/nm} \times \text{km)}}$$
$$D_{\text{Chromatic(ns/nm} \times \text{km)}} = -0.006 + 0.005$$
$$D_{\text{Chromatic(ns/nm} \times \text{km)}} = 0.001 \text{ ns/nm} \times \text{km}$$

$$T_{\text{Chromatic(ns)}} = D_{\text{Chromatic(ns/nm} \times \text{km)}} \times W_{\text{Spectral(nm)}} \times d_{\text{Install(km)}}$$
$$T_{\text{Chromatic(ns)}} = 0.001 \times 20 \times 50$$
$$T_{\text{Chromatic(ns)}} = 1.0 \text{ ns}$$

The optical fiber's total electrical bandwidth is as follows:

$$B_{\text{FiberTotal(GHz)}} = 0.35/T_{\text{Fiber(ns)}}$$
$$B_{\text{FiberTotal(GHz)}} = 0.35/1.0$$
$$B_{\text{FiberTotal(GHz)}} = 0.35 \text{ GHz}$$

6. The total fiber rise time $T_{\text{Chromatic(ns)}} = 1.0$ ns is less than the total equipment rise time $T_{\text{Equipment}} = 3.9$ ns, so the optical fiber selection is appropriate for the required equipment data transmission rate.

22.4 Network Topologies

Fiber optic networks should be configured to provide a flexible and versatile system that will allow the full benefits of the optical fiber to be realized. The traditional practice in fiber deployment is to install a dedicated fiber cable for each new application. This can be costly and can greatly limit the cable's potential. Systems should be installed to a carefully designed fiber optic cable deployment plan. Then, for example, when a single point-to-point application is installed, it can be upgraded to a future planned network.

Network topologies can be classified as *logical* or *physical topologies*. Logical topology describes the method by which the different nodes of the network communicate with each other. Physical topology is the actual physical layout of the cabling and nodes in the network (see Fig. 22.4).

22.4.1 Logical topologies

There are four standard logical topologies:

1. Point-to-point
2. Star
3. Bus
4. Ring

The following sections briefly describe each of these topologies.

1. Logical point-to-point. Logical point-to-point topology links two devices directly to each other (see Fig. 22.5). Common computer communication protocols use this topology, including RS232, RS422, V.35, T1, T3, and other proprietary protocols. Applications include computer modem connections, multiplexing links, two-way radio links, and satellite links.

2. Logical star. Logical star topology is an arrangement of point-to-point links that all have a common node (see Fig. 22.5). Applications include telephone PBX and multistation video monitoring systems.

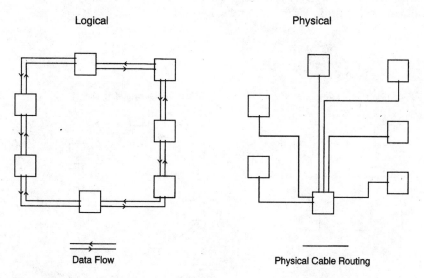

Figure 22.4 Logical and physical topologies.

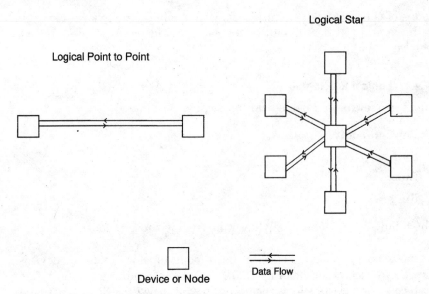

Logical Star

Logical Point to Point

Device or Node

Data Flow

Figure 22.5 Logical point-to-point and star topologies.

3. Logical bus. In logical bus topology, all devices are connected to one transmission bus, usually coaxial cable (see Fig. 22.6). On this bus, transmission occurs in both directions. When one device transmits information, all the other devices receive the information at the same time. This bus topology is the IEEE 802.3 and 802.4 standard. Applications include Ethernet and token bus.

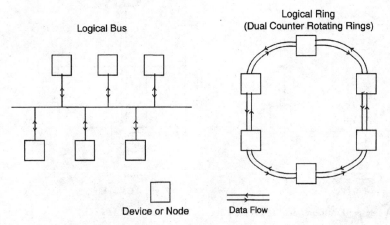

Logical Bus

Logical Ring
(Dual Counter Rotating Rings)

Device or Node   Data Flow

Figure 22.6 Logical bus and ring topologies.

4. Logical ring. Logical ring topology has all nodes connected in a ring (see Fig. 22.6). Transmission occurs on one cable in one direction in the ring. If two transmission rings are used, the arrangement is called a logical counter rotating ring, and transmissions move in opposite directions in each of the rings. This dual ring topology provides self-healing network protection in the event a cable or node should fail (see Fig. 22.7). Applications of this topology include token ring (IEEE 802.5) and FDDI (ANSI X3T9.5).

22.4.2 Physical topologies

The physical topology is medium-dependent and can be implemented in the same configuration as the logical topology. For example, a logical ring topology can be physically cabled to look like a ring. Each device is connected to the adjacent device in a physical ring configuration. Many networks are cabled with the same logical and physical topologies. For example, Ethernet 10Base5 is a logical bus topology and is often cabled as a physical bus topology. However, a logical topology can be connected in a different physical topology. The most common and recommended network physical topology (EIA 568)* is the star physical topology. It has a number of significant advantages over other physical topologies, including:

---

*Electronic Industries Association (EIA), 2001 Pennsylvania Avenue N.W., Washington, D.C.

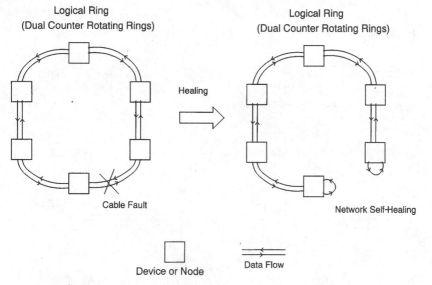

Figure 22.7 Counter rotating ring, self-healing.

- flexibility and ability to support many applications and all other logical topologies
- a centralized optical fiber cross connect location allowing easy fiber maintenance and administration
- recommended in EIA 568 standard for commercial building wiring
- many existing duct and conduit routes are often a star configuration
- easier system expansion

Two drawbacks of the physical star configuration include:

- a cable cut will cause the connected device or node to fail
- it requires more fiber optic cable length to implement than does a physical ring topology

Token ring and FDDI LAN networks are logical ring topologies, but they are often implemented as physical stars (see Fig. 22.8). A logical bus topology can be cabled as a physical star topology. The best example of this configuration is the Ethernet 10BaseT or 10BaseF network. Ethernet is a logical bus topology, however, the 10BaseT or 10BaseF cabling configuration is the physical star topology (see Fig. 22.9). The 10BaseF standard (IEEE 802.3 FOIRL, 10BaseFL standard)* is the optical fiber implementation. The concentrator acts as

---

*Institute of Electrical and Electronic Engineers, Inc. (IEEE), 445 Hoes Lane, P.O. Box 1331, Piscataway, NJ.

Logical Ring & Physical Ring   Logical Ring & Physical Star

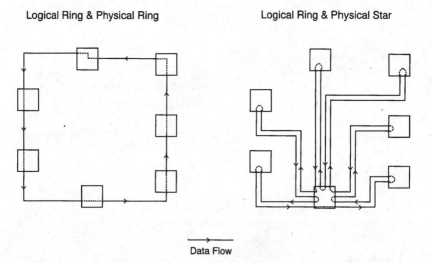

Data Flow

Figure 22.8 Logical ring implemented as a physical star.

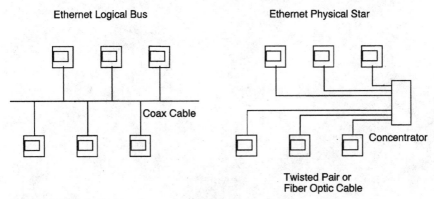

Figure 22.9 Bus logical topology as a physical star.

the lightwave equipment, converting the electrical signals to proper fiber optic signals for each workstation.

Fiber optic systems are usually deployed as point to point, star, or ring physical topologies. Point to point in common in modern applications. Star and ring are common in LAN networks. The ring physical topology is deployed when network protection is required in the event of a cable or node failure.

23

Personnel

An aspect often overlooked in many system implementations is the need for properly trained personnel to install, operate, and maintain the fiber optic system. Although fiber optic cable may look similar to electrical cable, it is very different. Many people fall into the trap of trying to "throw a fiber optic link into commission" with little or no knowledge of the subject. Although this approach may work because optical fiber is very forgiving, many problems can be encountered during and after the installation, and the full potential of the fiber may not be realized.

The best way to ensure that the system is designed and installed properly is to employ personnel trained in this field. Many companies provide this expertise or offer training to employees. The best instruction combines classroom fiber optic theory with practical, hands-on fiber optic system training. The practical, hands-on approach should include testing techniques that use a light source and power meter; fiber optic cable stripping, splicing, and connectorization; and optical fiber cleaning—all combined with proper safety concerns. The theory should remain practical (unless a design course is sought) and should include light transmission in optical fibers, fiber optic components, cable types and application, testing equipment and procedures, cable handling, installation, and maintenance and repair techniques, with safety concerns always emphasized. Training is also offered on an "in-house" basis. This allows employees to learn at the job site and become familiar with fiber optics in their work environment.

A good introductory practical course can be four full days in length, with additional days required to accommodate the hands-on training of each student. It is also recommended that for a first-time fiber optic installation a knowledgeable supervisor be at the job site to ensure it is done properly. Mistakes can easily occur and can be quite expensive to remedy.

The following is an example of a typical agenda for a basic fiber optic course:

Day 1

Introduction to Optical Fiber Communication
    History of optical fiber
    Advantages and drawbacks of optical fiber communication
    Practical examples of optical fiber communication
    Data, voice, and video communication over fiber
Introduction to Optical Fiber
    Optical fiber light spectrum
    Transmission techniques
    Analog signals, digital signals
Optical Fiber Composition
    Fiber diameters
    Light transmission in a fiber
    Multimode fiber
    Step index fiber
    Graded index fiber
    Single mode fiber
    Optical power loss
    Data rate limitations
    Optical fiber example
Fiber Optic Safety
    Safety precautions when working with optical fiber
Fiber Optic Cable Composition
    Loose tube and tight-buffered cable
    Cable jackets
    Strength members
    Fiber counts, cable size
    Cable specifications
    Cable procurement

Day 2

Method of Handling Fiber Optic Cable
    Bending radius
    Pulling tension
    General care
Outdoor Fiber Optic Cable Installation
    Duct/innerduct installation

Lubrication
Pulling tape
Buried cable installation
Cable pulling technique
Aerial installation
Indoor Fiber Optic Cable Installation
Conduit placement
Fire code rating
Riser installation
Equipment locations
Splicing and Termination Techniques
Splice enclosures, splice trays
Patch panel termination
Fusion splicing and mechanical splicing
Connector installation

Day 3

Cable Testing
Patch cords and connectors
Power meter testing techniques and dB
OTDR, fault locating
Installation test procedures
Acceptance criteria
Bit error rate test (BERT)
Receiver threshold test
System Integration
Equipment configuration
Multiplexer system
Modem system
Ethernet
FDDI
Light sources
Optical amplifiers
Receivers
General installation procedure
Maintenance and Repair
Maintenance
Cable repair
Records
Troubleshooting
Design basics

Day 4

Hands-on training for each student

Power meter testing techniques

Cable stripping

Mechanical splicing technique

Field connector installation technique

Each student assembles and tests a short fiber optic cable link.

In an advanced course, optical time domain reflectometer (OTDR) and fusion splicer techniques can be introduced.

Glossary of Terms and Acronyms

802.3 A local area network (LAN) protocol that uses CSMA/CD for medium-access control and a bus topology defined in Layers 1 and 2 of the OSI model.

802.4 A local area network (LAN) protocol known as token bus that uses a token-passing access method and a bus topology. Originated by General Motors and targeted for the manufacturing environment, it is defined in Layers 1 and 2 of the OSI model.

802.5 A local area network (LAN) protocol known as token ring uses a token-passing-with-priority-and-reservation access method and a star-wired ring topology. Originated by IBM, it is defined in Layers 1 and 2 of the OSI model.

Absorption The loss of optical power, in fiber, resulting from the conversion of light to heat. Causes include impurities, OH (water) migration, defects, and absorption bands.

Acceptance Angle The angle at which all incident light is totally and internally reflected by the optical fiber core. The acceptance angle is equal to sin NA and is also known as the *maximum coupling angle.*

Adapter A mechanical device used to align and join two fiber optic connectors. It is also referred to as the receptacle, coupling, or interconnecting sleeve.

ADM Add/drop multiplier.

Analog A waveform format that is continuous and smooth, used to indicate infinite levels of signal amplitude. See also *Digital.*

Angle of Incidence The angle between an incident ray and the normal to medium boundary.

Angle of Reflection The angle between a reflected ray and the normal to medium boundary.

Angle of Refraction   The angle between a refracted light ray and the normal to medium boundary.

Angstrom   A unit of length equal to 10^{-10} m or 0.1 nm.

ANSI   American National Standards Institute.

APD   Avalanche photodiode. A diode used for detecting very small quantities of light.

Aramid Yarn   This is a light material, usually yellow or orange, that provides strength and support to the fiber bundles in a cable. Kevlar is a type of aramid yarn that has very high strength.

Armor   Additional protection between cable jacket layers, usually consisting of corrugated steel.

Asynchronous   A signal that is not synchronized to a network clock.

ATM   Asynchronous transfer mode. A communication protocol standard using 53-byte packets defined in Layer 2 (data link) of the OSI model.

Attenuation, Optical Fiber   The diminution of light in the optical fiber. It is generally expressed without a negative sign in dB or dB/km. When specifying attenuation, it is important to note the applicable wavelength. Attenuation of an optical fiber is different for different optical wavelengths.

Attenuator, Optical   A device that reduces the fiber optic light beam intensity. Usually inserted at a connection point.

Backbone Cabling   The portion of telecommunications cabling that connects telecommunications closets, equipment rooms, buildings, or cities. It is a transmission medium (usually fiber optics) that provides a high-speed connection to numerous distributed facilities. It can be further classified as interbuilding or intrabuilding backbone cabling.

Bandwidth   The range of frequencies that a fiber can carry with minimum distortion. It is also the frequency at which the optical power is reduced by 50 percent (-3 dB) as a result of optical fiber dispersion effects and is expressed in megahertz $\times$ kilometer (MHz $\times$ km). It indicates the information-carrying capacity of an optical fiber or other transmission medium.

Bare Fiber Adapter   A bare fiber adapter is an optical fiber connector designed to temporarily connect an unterminated optical fiber to a connector. This allows for quick testing of unterminated fibers.

Baud Rate   The number of electrical transitions of a transmitted digital signal per second. Not the same as data rate. A modem can have a transmission data rate higher than its baud rate.

Bellcore   Bell Communications Research. A research and development organization owned by a number of American Bell operating companies that develop communication standards.

Bit   One binary digit.

Bit Error Rate (BER)   The ratio of bits received in error to the number of bits sent. A BER of 10^{-9} is common (a billion bits sent with one bit received in error).

Bit Error Rate Tester (BERT) An instrument that measures the amount of bits transmitted incorrectly by a digital communication system.

bps Bits per second.

Bridge Connects two or more similar LANs in Layer 2 of the OSI model.

Broadband Data rates at or greater than 45 Mbps (or DS3).

Brouter A vendor device that acts as both a bridge and a router.

Buffer A protective cover of plastic or other material, usually color-coded, that covers the optical fiber. A buffer can either be tight, as in a tight-buffered cable adhering directly to the optical fiber coating, or it can be loose, as in a loose tube cable, where one or more fibers lie loose in the buffer tube. The buffer must be stripped off for cleaving and splicing.

Byte Eight adjacent binary digits (bits).

CAD Computer-aided design.

CAM Computer-aided manufacturing.

CATV Community antenna television, also known as cable television system.

CCITT Consultative Committee on International Telephone and Telegraph. An international committee that develops and recommends standards for telecommunications.

CCTV Closed-circuit television.

CDDI Copper Distributed Data Interface. A protocol standard similar to FDDI but that uses unshielded twisted pairs or shielded twisted pairs to provide 100 Mbps data communications.

Channel A communication path that is normally fully duplex.

Cladding A layer of glass or other material with a low index of refraction that surrounds the fiber core, causing the light to stay captive in the core.

Cleaver An instrument used to cut optical fibers in such a way that the ends can be connected with low loss.

Cluster Controller IBM control unit models 3174, 3274, and so forth, which can simultaneously service IBM 3178, 3278, or other terminals.

CO Central office. A local telephone company office that contains a switch for terminating subscribers.

Coating A thin layer of plastic, or other material, usually 250 or 500 μm in diameter and color-coded, that covers the cladding of a fiber. Most fibers have a coating that must be stripped for cleaving and splicing.

Codec A device that converts analog signals to digital signals or vice versa.

Concentrator An electronic device used in LANs that allows a number of stations to be connected to a single data trunk.

Conductor A material that allows the flow of electric current.

Conduit   A pipe or tube where cables can be housed.

Connector (fiber optic)   A device that joins two optical fibers together in a repeatable, low-optical-loss manner.

Cord   A short length of flexible cable used for connecting equipment.

Core   The inner portion of the fiber that carries light. Light stays in the core due to the difference in refractive index between the core and cladding.

Coupler (fiber optic)   A device that joins together three or more optical fiber ends so that an optical signal can be split, or transmitted, from one fiber to two or more fibers.

CRC   Cyclic redundancy check. A mathematical checking method for determining the integrity of a data packet.

Critical angle   The smallest angle at which a meridional light ray can be totally reflected in a fiber core.

Cross-Connect, Optical   In fiber optics, a patch panel is used as a cross-connect.

CSA   Canadian Standards Association.

CSMA/CD   Carrier Sense Multiple Access with Collision Detection. An access protocol used by Ethernet/802.3.

CSU   Channel service unit. A device that allows testing functions on a channel such as loop back.

Cutoff Wavelength   The shortest wavelength the fiber will propagate; primarily specified in single mode fibers.

Dark Current   The current that flows in a photodetector when there is no light on the detector.

Dark Fiber   Optical fibers, in a cable, that have no fiber optic modules connected to them (usually spare fibers). These are offered by some carriers, allowing the user to install his or her own optical transmission equipment.

Data Rate   The number of bits of information that can be transmitted per second. Expressed as Gbps, Mbps, kbps, or bps.

dB   Decibel.   A logarithmic measure of optical power.

dBm   Decibel relative to 1 milliwatt; dBm = 10 log (output power milliwatts/1 milliwatt).

DCE   Data communication equipment. Communication equipment used as the interface to a communications channel for data terminal equipment (DTE) such as a modem.

Dead Zone Fiber   A length of optical fiber (usually 1 km), connected between the OTDR and fiber facility, used to move the fiber facility OTDR trace beyond the OTDR's dead zone.

Dielectric   A material that will not conduct electricity (insulator).

Digital   A data waveform format that has only two physical levels, corresponding to zeros and ones. See also *Analog*.

Dispersion Distortion of a light pulse caused by the propagation characteristics of different wavelengths and different mode paths.

Dispersion Flattened Fiber An optical fiber that is optimized for minimum dispersion at both 1310 nm and 1550 nm.

Dispersion Shift Fiber An optical fiber that is optimized for minimum dispersion of 1310 nm or 1550 nm.

DNA Digital Network Architecture.

Double Window Fiber A fiber designed to operate at two different wavelengths.

DS0 Digital Signal Level 0. A digital communication channel that is 64 kbps. Twenty-four DS0 makes one DS1 channel (T1).

DS1 Digital Signal Level 1. A digital communication channel that is 1.544 Mbps, T1 rate.

DS2 Digital Signal Level 2. A digital communication channel that is 6.312 Mbps, T2 rate.

DS3 Digital Signal Level 3. A digital communication channel that is 44.736 Mbps. It consists of 28 DS1.

DSU Digital Service Unit. A physical layer device that provides digital channel testing and monitoring functions.

DTE Data terminal equipment. User terminal equipment such as a computer, terminal, or workstation.

DTMF Dual tone multiple frequency, also known as touchtone. A set of audio frequencies used in telephone signaling. Often used by telephone sets.

E1 A European data communication standard rate of 2.048 Mbps, which can carry thirty 64 kbps channels.

E3 A European data communication standard rate of 34.368 Mbps, which can carry sixteen E1 channels.

EIA Electronic Industries Association.

EMI Electromagnetic interference.

Ethernet A 802.3 LAN technology that uses CSMA/CD access method and a bus topology.

Expansion Loop A loop placed in a supported cable to compensate for the expansion and/or contraction of the supporting member and/or the cable.

FDDI Fiber distributed data interface. This is a fiber optic networking technology particularly suited for high data rate traffic. It uses a token-passing method and a dual counter rotating ring topology at a data rate of 100 Mbps.

Ferrule The rigid center portion of a fiber optic connector, usually made of steel or ceramic.

Fiber Bandwidth The transmission frequency at which the signal magnitude decreases to half of its optical power (-3 dB).

Fiber Count The number of optical fibers between any two locations. It can also refer to the number of optical fibers in a single cable.

Fiber Optics The transmission of light through optical fibers for communication and signaling.

Flame-Test Rating (FT 1, 4, 6) A Canadian flame-test rating for cables established by the Canadian Standards Association.

FOT Fiber optic terminal.

Frequency The number of repetitions of a periodic action per unit of time. Units of frequency are Hertz (Hz).

Fresnel Reflection Reflection of a portion of light at a boundary between two materials having different indexes of refraction.

FT1 Fractional T1. A fraction of the full 24 channels in a T1 carrier.

FTTC Fiber to the curb. The fiber optic architecture wherein fiber optics is deployed most of the way to the customer's home, but stops at the curb. From the curb, copper pairs or coaxial cable are used to bring the signal into the home.

FTTH Fiber to the home. The fiber optic architecture wherein fiber optics is deployed directly into the customer's home.

Full Duplex The ability to transmit and receive signals at the same time.

Full-Width Half Maximum (FWHM) FWHM is the spectral width in nanometers (nm) of an optical source at one-half peak optical power level.

Fusion Splice A permanent joint of two optical fibers performed by the precise melting of two fibers together.

giga- The prefix meaning one billion (10^9).

Graded Index Fiber An optical fiber in which the index of refraction of the core gradually decreases as you move toward the cladding.

Ground A common return point to the earth for electrical current, usually through a ground rod.

Ground Loop Currents Undesirable ground currents that cause interference; usually created when grounds are connected at more than one point.

Half Duplex A communication method in which one end must wait until a transmission is complete before information is sent and vice versa: transmission and reception of communications must alternate and cannot occur at the same time.

HDTV High-definition television.

Hertz (Hz) One cycle per second.

Horizontal Cabling (or Wiring) The portion of cabling that provides connectivity between telecommunication closet and work area telecommunication outlet.

Hub A communication device that uses a star-wiring pattern topology; common in LANs.

Hybrid Fiber/Coax (HFC)   A cable that contains both optical fiber and coaxial cable.

IEEE   Institute of Electrical and Electronics Engineers.

Impedance   The total opposition an electric circuit offers to an alternating current flow that includes both resistance and reactance.

Index Matching Fluid   A liquid or gel with a refraction index that matches the core of the fiber.

Index of Refraction   The ratio of the speed of light in a vacuum to the speed of light in a material. Air $n = 1.003$; glass $n = 1.4$ to 1.6. N also varies slightly for different wavelengths.

Insertion Loss   The optical power lost due to the insertion of an optical component such as a connector, splice, or attenuator.

Insulator   A material that does not conduct electricity under normal operating conditions.

Interbuilding Cabling   Cabling between buildings.

Intermediate Cross-Connect (IC)   A secondary cross-connect in backbone cabling used to provide connection to telecommunication closets.

Intra-building Cabling   Cabling within a building.

ISO   International Organization for Standards.

Jacket   The outer coating of a wire or cable.

kilo-   The prefix meaning one thousand (10^3).

LAN   Local area network. A network in a local area, such as in a single large building.

Laser   An acronym standing for "light amplification by stimulated emission of radiation." A device that produces light by the stimulated-emission method, with a narrow range of wavelengths, which are emitted in a directional coherent beam. Most fiber optic lasers are solid-state devices.

LED   An acronym standing for "light-emitting diode." A semiconductor device that produces incoherent light when current passes through it.

Light, Fiber Optic   The spectrum of light at 850-nm, 1310-nm, or 1550-nm wavelength.

Light Ray   The direction the light waves travel.

Lightwave Equipment   Any electronic communication equipment that is used for optical fiber communication; also known as optical terminating equipment or optical modems.

M13   Multiplexer DS1 to DS3.

Main Cross-Connect (MC)   The central main cross-connect in backbone cabling used to provide connectivity between intermediate cross-connects.

MAN   Metropolitan area network. A network connecting multiple sites in one geographical area.

Maximum Coupling Angle The angle at which all incident light is totally and internally reflected by the optical fiber core. The maximum coupling angle is equal to sin NA.

Mechanical Splice A small device that joins and secures two fibers together using mechanical means.

mega- A prefix meaning one million (10^6).

micro- A prefix meaning one millionth (10^{-6}).

Microbending Loss Loss in a fiber caused by sharp curves of the core with displacements of a few micrometers. Such bends may be caused by the buffer, jacket, packaging, installation, and so on. Losses can be significant over a distance.

milli- A prefix meaning one thousandth (10^{-3}).

Minimum Bend Radius The minimum radius of a curve that a fiber optic cable or optical fiber can be bent on without any adverse affects to the cable's or optical fiber's characteristics.

Modal Dispersion *See* Multimode dispersion.

Mode A single electromagnetic light wave that satisfies Maxwell's equations and the boundary conditions given by the fiber. It can be simply considered as a light ray path in a fiber.

Mode Field Diameter (MFD) The diameter of the light spot from a single mode fiber. It is usually not the same diameter as the core diameter.

Mode Scrambler A device that causes modes to mix in a fiber.

Modem An electronic device that converts one form of signal to another form using a modulation technique. An optical modem converts the electrical signal to an optical signal and vice versa.

Modulation Altering a carrier wave so that it can carry signal information. Optical modulation involves altering the lightwave amplitude or frequency to convey signal information.

Multimode Dispersion (Modal Dispersion) The distortion of a light pulse or light wave caused by light rays traveling different paths in a fiber, and consequently arriving at the destination at different times.

Multimode Fiber A fiber that propagates more than one mode of light.

Multiplexer An electronic unit that combines two or more communication channels onto one aggregate channel. It can be further classified into time domain or frequency domain multiplexer.

nano- A prefix meaning one billionth (10^{-9}).

NEC National Electrical Code.

NEMA National Electrical Manufacturers Association.

Nonreturn to Zero (NRZ) A digital code in which the signal level is high for a 1 bit, low for a 0 bit, and does not return to low for successive 1 bits.

Numerical Aperture (NA) Numerical aperture is the sine of the angle measured between an incident light ray and the boundary axis that an optical fiber can accept and propagate light rays. It is a measure of the light-accepting property of the optical fiber.

OC-1 to OC-48 Protocol channels for Sonet. Rates range from 51.84 Mbps OC-1 to 2.488 Gbps OC-48. Defined in the Physical Layer 1 of the OSI model.

Ohm The electrical unit of resistance.

Optical Dynamic Range The receiver's optical dynamic range is the light-level window in dBm at which a receiver can accept optical power.

Optical Fiber A single optical transmission element comprised of a core, cladding, and coating.

Optical Loss The reduction of fiber optic power caused by a substance.

Optical Margin The allowance for optical loss that is in addition to the total of all optical losses in a fiber optic link.

Optical Source An electronic unit used to convert electrical signals to corresponding optical signals for transmission in an optical fiber.

OSI Model Open Systems Interconnection (OSI) seven-layer protocol model developed by the International Standards Association.

OTDR Optical time domain reflectometer is a test instrument that sends short light pulses down an optical fiber to determine the fiber's characteristics, attenuation, and length.

Packet A grouping of data that has address header and control information.

Patch Cord, Optical A short length of thick-buffered optical fiber with connectors at both ends. It can be used to connect equipment together, connect equipment to patch panels, or connect jumper patch panels.

Patch Panel A group of connection points used to mechanically terminate a cable and provide connection to individual fiber optic patch cords.

PBX Private branch exchange.

Photodetector A device that converts light energy into electrical energy. A silicon photodiode is commonly used in fiber optics.

pico- A prefix meaning one trillionth (10^{-12}).

Pigtail A short length of buffered optical fiber with a connector at one end. It is used to terminate fibers in a cable.

Plenum A space that can be found above dropped ceilings or in raised floors that is used to circulate air. Also a fire code rating for cable.

PLC Programmable logic controller.

Polyethylene A thermoplastic material often used for cable jackets.

Polyvinyl Chloride A thermoplastic material often used for cable jackets.

POP Point of presence. A physical location at which a carrier provides services to a customer.

POTS Plain old telephone system. A two-wire loop start ring telephone connection with touchtone or make-break signaling.

Protocol A set of rules that will enable data communications.

Pull Box An enclosure placed in a conduit or a duct route to allow for access to cable during cable installation or removal operations.

Pulling Tension The tension on a cable during cable pulling installation operations.

Rayleigh Scattering The scattering of light (or loss of light) due to variation in medium optical density, composition, and molecular structure.

Receiver, Optical An electronic unit that converts light signals to electrical signals.

Reflection The reflection of a light ray at a boundary, of two dissimilar mediums, back into the first medium.

Refraction The change of direction and velocity of a light ray through a boundary of two dissimilar mediums.

Repeater A device that repeats and regenerates a signal.

Return Loss The amount of fiber optic light reflected back to the source.

Return to Zero (RZ) A digital code in which the signal level is high for a 1 bit for the first half of the bit interval and then goes to low for the second half of the bit interval. The level stays low for a 0-bit complete interval.

Riser A vertical shaft or space between floors in buildings used to route cables. Also a fire code rating for cables.

Router A LAN device that operates at the OSI model network layer and is used to connect dissimilar LANs or networks together.

Sensitivity The minimum amount of optical power that the lightwave equipment needs to receive to achieve signal transmission to equipment specifications.

Single Mode Fiber A fiber that carries only one mode of light. Only one light path is propagated.

Slit Duct A duct that is slit along its length. This allows cable placement into the duct without pulling the cable into the duct.

SNA Systems network architecture. IBM's seven-layer data communication architecture.

Sonet Synchronous optical network. A protocol standard for communication over optical fiber defined in Layer 1 (physical layer) of the OSI model.

Spectral Bandwidth The spectrum of frequencies in which the intensity of light is greater than half its peak intensity.

Speed of Light 2.998×10^8 meters per second in a complete vacuum but less in any other material.

Splice A permanent junction between two fibers created by the fusion splice method or by the mechanical splice method.

Splice Enclosure An enclosure used to house fiber optic splices and splice trays.

Step Index Fiber An optical fiber in which the core has a constant index of refraction.

STP Shielded twisted-pair wire.

Stratum A primary reference clock used for network synchronization.

Supported Radius The radius of a duct or conduit bend when it is continuously supported around its bend.

Synchronous A signal that is synchronized to a network clock.

Synchronous Transmission Data communication protocol in which the data is sent continuously.

T1 A communications aggregate of a data rate of 1.544 Mbps that has 24 channels.

Telecommunication Closet (TC) The facility used to provide connection between backbone cabling and horizontal cabling.

Telecommunication Outlet The horizontal cabling terminal device located in a work area used to connect termination equipment such as computers.

Tight-Buffered Cable A fiber optic cable where each fiber has a 900 μm plastic coating.

Total Dispersion The combined total of chromatic and multimode dispersion.

Total Internal Reflection All light that is incident on a medium boundary is reflected back into the first medium.

Transmitter, Optical An electronic unit that converts electrical signals to light signals.

UL Underwriters Laboratories.

Unsupported Radius The radius of a duct or conduit when it is not continuously supported around its bend.

UTP Unshielded twisted-pair wire.

WAN Wide area network. A network that extends into different cities.

Wavelength Division Multiplexing, Optical The combining of two or more optical signals of different wavelengths onto one fiber.

B

Units

Physical constants

Speed of light in vacuum:	$c = 2.998 \times 10^8$ m/s
Planck's constant:	$h = 6.63 \times 10^{-34}$ joule $\times$ s
Boltzmann's constant:	$k = 1.38 \times 10^{-23}$ joule/kg
Electron charge:	$e = -1.6 \times 10^{-19}$ C

Prefixes used to indicate multiples

Prefix	Factor	Symbol
exa-	10^{18}	E
peta-	10^{15}	P
tera-	10^{12}	T
giga-	10^{9}	G
mega-	10^{6}	M
kilo-	10^{3}	k
hecto-	10^{2}	h
deka-	10	da
deci-	10^{-1}	d
centi-	10^{-2}	c
milli-	10^{-3}	m
micro-	10^{-6}	μ
nano-	10^{-9}	n
pico-	10^{-12}	p
femto-	10^{-15}	f
atto-	10^{-18}	a

Units

Unit	Symbol	Measurement
meter	m	length
gram	g	mass
volt	V	voltage
ohm	Ω	resistance
ampere	A	current
coulomb	C	charge
watt	W	power
joule	J	energy
farad	F	capacitance
second	s	time
Celsius	°C	temperature
kelvin	K	temperature
bit	b	pulse
byte	b	8 bits
second	s	time
Hertz	Hz	frequency
decibel (mW)	dBm	power

Digital transmission rates

Designation	Data rate	Voice channels (64 kbps)
DS0	64 kbps	1
FT1	(1 to 24) × 64 kbps	1 to 24
T1–DS1	1.544 Mbps	24
T2–DS2	6.312 Mbps	96
T3–DS3	44.736 Mbps	672
T4–DS4	274.175 Mbps	4032

Sonet transmission rates

Designation	Data rate	Voice channels (64 kbps)
OC-1	51.84 Mbps	672
OC-3	155.52 Mbps	2016
OC-12	622.08 Mbps	8064
OC-24	1244.16 Mbps	16,128
OC-48	2488.32 Mbps	32,256

Optical Fiber Color Codes

Standard EIA/TIA 598 Color Code*

Fiber number	Tube color	Fiber color
1	Blue	Blue
2	Blue	Orange
3	Blue	Green
4	Blue	Brown
5	Blue	Slate
6	Blue	White
7	Blue	Red
8	Blue	Black
9	Blue	Yellow
10	Blue	Violet
11	Blue	Rose
12	Blue	Aqua
13	Orange	Blue
14	Orange	Orange
15	Orange	Green
16	Orange	Brown
17	Orange	Slate
18	Orange	White
19	Orange	Red
20	Orange	Black
21	Orange	Yellow
22	Orange	Violet
23	Orange	Rose
24	Orange	Aqua
25	Green	Blue
26	Green	Orange
27	Green	Green
28	Green	Brown
29	Green	Slate
30	Green	White
31	Green	Red
32	Green	Black

Source: Electronic Industries Association, 2001 Pennsylvania Ave., NW, Washington D.C.

Other common color codes

Cable Manufacturer A

Fiber number	Tube color	Fiber color
1	Black	Blue
2	Black	Orange
3	Black	Green
4	Black	Brown
5	Black	Slate
6	Black	White
7	Red	Blue
8	Red	Orange
9	Red	Green
10	Red	Brown
11	Red	Slate
12	Red	White
13	Yellow	Blue
14	Yellow	Orange
15	Yellow	Green
16	Yellow	Brown
17	Yellow	Slate
18	Yellow	White
19	Violet	Blue
20	Violet	Orange
21	Violet	Green
22	Violet	Brown
23	Violet	Slate
24	Violet	White

Cable Manufacturer B

Fiber number	Tube color	Fiber color
1	Red	Brown
2	Red	Blue
3	Red	Orange
4	Red	White
5	Red	Slate
6	Red	Green
7	Black	Brown
8	Black	Blue
9	Black	Orange
10	Black	White
11	Black	Slate
12	Black	Green
13	Yellow	Brown
14	Yellow	Blue
15	Yellow	Orange
16	Yellow	White
17	Yellow	Slate
18	Yellow	Green
19	Green	Brown
20	Green	Blue
21	Green	Orange
22	Green	White
23	Green	Slate
24	Green	Green

Cable Manufacturer C

Fiber number	Tube color	Fiber color
1	Blue	Blue
2	Blue	Orange
3	Blue	Green
4	Blue	Brown
5	Blue	Slate
6	Blue	White
7	Orange	Blue
8	Orange	Orange
9	Orange	Green
10	Orange	Brown
11	Orange	Slate
12	Orange	White
13	Green	Blue
14	Green	Orange
15	Green	Green
16	Green	Brown
17	Green	Slate
18	Green	White

Fiber Optic Records

Optical Fiber Attenuation Test Results

Date _____
Cable Location _____
Location A _____
Location B _____

Tested by _____
Cable Length (OTDR) _____ (Physical) _____ Factor _____
Fiber Type _____
Cable Manufacturer _____

Fiber Color _____

| Bundle Color/Fiber Color | Test Wavelength _____ nm | | | | Test Wavelength _____ nm | | | |
| | OTDR Measurements | | | Power Meter (dB) Loss | OTDR Measurements | | | Power Meter (dB) Loss |
	From Location A (dB) Loss	From Location B (dB) Loss	Average Loss (dB) (A + B)/2		From Location A (dB) Loss	From Location B (dB) Loss	Average Loss (dB) (A + B)/2	
1.								
2.								
3.								
4.								
5.								
6.								
7.								
8.								
9.								
10.								
11.								
12.								
13.								
14.								
15.								
16.								

Optical Splice and Anomaly Test Results

Date _____

Tested by _____

Cable Location _____

Cable Length (OTDR) _____ (Physical) _____ Factor _____

Location A _____

Fiber Type _____

Location B _____

Cable Manufacturer _____

Test Wavelength _____ nm

Test Wavelength _____ nm

Fiber Color	OTDR Measurements of Splice or Anomaly			Location of Splice or Anomaly	OTDR Measurements of Splice or Anomaly			Location of Splice or Anomaly
Bundle Color/Fiber Color	From Location A (dB) Loss	From Location B (dB) Loss	Average Loss (dB) (A + B)/2		From Location A (dB) Loss	From Location B (dB) Loss	Average Loss (dB) (A + B)/2	

Equipment Optical Power Output Record

Date _____ Tested by _____

Location _____

Power Meter Model _____

Equipment	Location	Test Pattern	Output Power in dBm

Optical Fiber Return Loss Record (Single Mode Fiber Only)

Date _____ Tested by _____

Location _____

Return Power Meter Model _____

Fiber Number	Panel	Return Loss dB

Optical Link Budget

Optical Fiber Diameter: Core_____ μm/Cladding_____ μm

Optical Fiber Numerical Aperture (NA):_____

Lightwave Equipment Wavelength: _____ nm

a.   Optical Fiber Loss at Operating Wavelength: _____
       _____ km length at_____ dB/km   _____ dB

b.   Splice Loss:
       _____ splices × _____ dB/splice   _____ dB

c.   Connection Loss:
       _____ connections ×_____ dB/connection   _____ dB

d.   Other Component Losses: _____ dB

e.   Design Margin: _____ dB

           _____

f.   **Total Link Loss**   _____ **dB**
         (a + b + c + d + e)

g.   Transmitter Average Output Power: _____ dBm

h.   **Receiver Input Power**   _____ **dBm**
         (g − f)

i.   Receiver Dynamic Range: _____ dBm to_____ dBm

j.   Receiver Sensitivity for (BER or S/N) _____ :   _____ dBm

k.   **Remaining Margin:**   _____ **dB**
         (h − j)

Notes:

1. Remaining margin, line k, must be greater than zero for proper system design.

2. In line c, exclude lightwave equipment optical fiber connection loss.

3. Receiver input power, line h, must be within receiver dynamic range, line i, for proper operation.

Index

Note: The *f*. after a page number refers to a figure.

ABOUT THE AUTHOR

Bob Chomycz, P. Eng., is founder of Telecom Engineering Inc., a company that specializes in fiber optic telecommunication system engineering, implementation, and training. With over 10 years of experience as a telecommunications engineer, Bob has accumulated a wealth of knowledge in fiber optics which he shares with you. His experience includes engineering design, integration, installation, and commissioning a large variety fiber optic project for data, LAN, WAN, voice, and video communications. At present, he is involved in helping industry build state-of-the-art fiber optic communication systems.